THE ROMANCE OF MATHEMATICS

THE ROMANCE OF MATHEMATICS

BEING

THE ORIGINAL RESEARCHES

OF

A Lady Professor of Girtham College

IN

Polemical Science, with some Account of the Social Properties of a Conic; Equations to Brain Waves; Social Forces; and the Laws of Political Motion

P. Hampson, M.A.

ORIEL COLLEGE, OXFORD

Preface by

Denny Taylor

Professor Emeritus, Literacy Studies

NEW YORK, NY

Published by Garn Press, LLC
New York, NY
www.garnpress.com

ISBN: 978-1-942146-14-8 (paperback)
ISBN: 978-1-942146-15-5 (hardcover)

Contents

Master Your Heart!

Girls Are As Good At Mathematics As The Boys In Their Class

To every young woman who believes she is no good at mathematics I say, "Master your heart!" The very idea that you are bad at math has been inculcated in you through the nefarious practices of the gender that is often called "your better half!"

"Ah, you male sycophants!" P. Hampson tells us a Lady Professor of Girtham College writes in her original researchers, *The Romance of Mathematics.*

Hampson explains he found her learned papers on mathematics – described by him as "original thoughts and investigations" – in the drawer of her well-worn desk at Cambridge University, and he subsequently published them in 1886. Elliot Stock was the publisher.

Hampson writes of the "mysterious desk" and describes the papers as "profound treasures". He states, "her words excite". The essays focus on "female education", "Cambridge Man's Powers of Application", the "Social Properties of a Conic Section", "Polemical Mathematics", "Circles" and the

"Eccentricity of Curves".

Other topics on which the Lady Professor writes include, "Alarming Result of Suppression", "Directrix", "Contact of Curves and States", "Radical Axis", "Attractive Forces", "Impulsive Forces", "Friction" and the "First Law of Motion".

Finally, Hampson tells us that her original researches end with – what else? – nothing, other than "Marriage".

"You would prevent us from competing with you," Hampson is almost laughing on the page when he writes of the way in which the Lady Professor scolds her male counterparts.

"You would separate yourselves on your island of knowledge, and sink the punt which would bear us over to your privileged shore," Hampson states she writes, "Of all the twaddle – forgive me, male sycophants! – that the world has ever heard, I think the greatest is that which you have talked about female education."

"Now," according to Hampson, she writes, "look me straight in the face (no shirking, sir!). Is it not jealousy – green-eyed, false-tongued jealousy – which saps your generous instincts, and makes you talk rubbish and nonsense about the strains, and brains, and ambition, and the like? And if that is not hypocritical, I do not know what is."

I can imagine you might be surprised at the ferocity of our Lady Professor's female challenge to the superiority of men at mathematics, especially when you consider your own experiences, your sweaty-hands, and the heart-bumping, stomach-retching effects of the proctored math tests that you have taken.

"Sometimes (it) takes my breath away even to read the terrible excruciating things," Hampson tells us our Lady Professor writes, "which seem to turn one's brain round and round, and contort the muscles of one's face, and stop the pulsation of one's heart, when one tries to grasp the horrid things."

Hampson was not intimidated by our Lady Professor, nor was he surprised by her accusations of the trumped-up superiority of society's sycophantic men. How could he be? After all, our Lady Professor of Girtham College in Polemical Sciences was strumped-up by the ink of his pen.

The Romance of Mathematics is a 19th century skit, a satirical work in keeping with the literary vernacular of the day that became an Oxford and Cambridge tradition. In the 20th century, Alan Bennett, Graham Chapman, John Cleese, Peter Cook, Dudley Moore, and Michael Palin carried on the Oxbridge tradition, turning skits into mass media sensations, including *The Goon Show*, *Beyond the Fringe*, *Monty Python*, and *That Was the Week that Was* – all include biting satires on class, gender, and race.

In the journal Nature, May 10, 1888 (vol. 38, pp. 28-29), a review of *The Romance of Mathematics* begins with the statement: "We had no idea that the task before us was to examine and report upon a somewhat mild *jeu d'esprit*." The review in the learned scientific journal ends:

> We ought to apologize for going into such detail, but our account will show our readers that the present work does not deal with mathematical discoveries. It is a "skit" with the perusal of which a reader acquainted with mathematics may while away, not unpleasantly, and odd half-hour or two.

So let's while away the time if only for a minute and examine the content of this mild *jeu d'esprit*. Our male writer, masquerading as a female mathematician, writes:

> A uniform rod ... of length 2c ... whose plane is vertical (why not incline it 30°?) ... passes through a small,

smooth, fixed ring situated in the axis at a distance b from the vertex ... (so) that if the equilibrium be slightly disturbed, the rod will perform small oscillations with its lower end on the arc of the cycloid in the time

$$4\pi \sqrt{\dfrac{a\{c^2 + 3(b - c)^2\}}{3g(b^2 - 4ac)}},$$

where 2a is the length of the axis of the cycloid.

"A sweet pretty problem, truly," our masquerader writes, interjecting himself into the sexual act in this seemingly benign work of deception. He tells the reader that he has found a photograph of this "accomplished authoress".

Hampson waxes lyrical about the Lady Professor and her "considerable beauty". He describes her eyes as "clear, steadfast, and open as the day." He writes about the "firmness about her mouth", then quickly adds, "but it is a sweet and pretty one notwithstanding; and a smile, half scornful, half playful, can be detected lurking about the corners of the lips, which do not seem altogether fitted for pronouncing hard mathematical terms and abstruse scientific problems."

And there you have it. A woman's lot is to be man's *jeu d'esprit*, a beautiful sexual object whose capacity for mathematical thought and scientific thinking is not to be taken seriously. I will argue differently that, when the curtain is drawn back and the lights go up, this satirical skit is nothing more than a manifestation of gendered intolerance for the idea that women are men's equal at mathematics – *if* they are given the opportunity to participate in this male dominated activity.

But first, who was P. Hampson?

P. Hampson, the writer of this clever work of fiction, was in fact Peter Hampson Ditchfield. At first he seems like an unlikely writer for this footlights beyond-the-fringe skit. He was born in 1854 and educated at Oriel College, Oxford, where he studied mathematics. He received a third-class B.A. degree. He was ordained in 1878 and became a priest the following year. His obituary states that he "occupied a distinguished place in the literary life of the country" and that his "easy, graceful style" made many of his books "a very great vogue".

The Romance of Mathematics, published in 1886, was his second book, and it appears in a long list in his obituary. His books have such titles as *Our English Villages*, *Cathedrals of Great Britain*, and the *Handbook of Gothic Architecture*. His *jeu d'esprit* skit was certainly not his magnum opus. It would seem an unlikely book for Ditchfield to have written, if it were not for *The Sorceress of Paris*, which he published in 1896. The romance novel is described as the "chronicle of Jean Louis Charles Count du Dunois" in which Susanne, the sorceress, has occult powers, including second sight. She is described in one account of the book as "a white witch who pays dearly for the use of her gift."

There is also the fact that the Rev. Ditchfield was a prominent Freemason, who was referred to in one publication as "*Bro* Ditchfield" who practiced the "Masonic craft". Initiated in 1893 and a master, founder, or honorary member of plus or minus a dozen Masonic lodges, he became the *Grand Chaplain of the Freemasons of England* in 1917.

It does not take much analysis to reach the conclusion that the writer of *The Romance of Mathematics*, the Reverend Ditchfield, *Bro* Ditchfield to his rich and illustrious Masonic friends, was fortunate to live a superior life – a male dominated existence in which both gender and class had preordained his elevated social place in British society since birth.

Nothing much has changed. In the UK as in the US, social inequality is as great now as it was in the 19th century. Mathematics, physics, chemistry, and the philosophical sciences are still dominated by men, and so is the Christian church. And the Freemason is still a secret society that, while proffering to be more inclusive, is still the bastion of rich white men who hold positions of power in a global society – that has yet to recognize the mathematical abilities of women are equal to the mathematical abilities of men.

In this regard I offer the following example to men and women who might read this book, and who take it upon themselves to challenge the mores of these times, which have been instilled in them since birth.

It is more than 50 years since I was given special permission to take mathematics with the boys. I was 14 years of age and the boys had been studying "maths" since they were 11 years old. In my school, girls did not take mathematics. They took arithmetic instead.

I was 13 by the time the maths master – we called him "Sir" – recognized that I was "good at maths", but it took him another year to persuade the needlework mistress and the domestic science mistress to release me from their classes for half a year so I could study mathematics with the boys in my class.

The maths master proposed I take maths instead of needlework in the autumn term and maths instead of domestic science the following term. Thus, given that there were three terms, the first year I would miss two terms of domestic science and the second year (my last in the school) I would miss two terms of needlework.

Neither the needlework mistress nor the domestic science mistress was happy about this arrangement, and they argued most strenuously with the mathematics master for an entire year. The irony of this situation was not lost on me. Two educated women, so enculturated into the importance of my being able to sew and cook that they couldn't even entertain the idea that it might be more advantageous for me to study mathematics. In contrast, the mathematics master thought it completely unacceptable that I was being taught to make buttonholes and darn socks (literally) instead of studying algebra and geometry.

Then, at the beginning of the autumn term of my fourteenth year, the mathematics master instructed me to attend maths classes with the boys, and I did just that.

On the first day that I was to take maths, the needlework mistress walked into the classroom of the mathematics master to collect me and escort me to her needlework class. The mathematics master told her to leave his classroom and, with his arms out to guide her, he followed her out as she began to shout. He shouted back. The classroom was a prefabricated unit, and the boys in the class got up from their seats and watched through the windows as the needlework mistress walked up and down yelling, while the mathematics master held his ground, blocking the door to the classroom where I sat in my seat and waited.

Needless to say, for the boys it was a no-win situation. They wanted the mathematics master to win the fight, but they also wanted me to leave, and so when he walked back in they were totally silent as they awaited the decision.

Without looking at me the mathematics master walked to the board, picked up some chalk, and began the lesson.

For the rest of the term the boys shunned me in class when Sir was present, and did other things that were not very pleasant when he stepped out of the classroom. And it was clear to me from the very first class that if I was to sur-vive, I would have to prove I could earn my elevated status. The boys needed to know my capacity for mathematics was no less than theirs, and the mathematics master had to be able to tell the needlework mistress that I had passed the end-of-term exams.

At the end of my first term I placed first, and there was a large gap between my score and the score of the boy who placed second. Some of the boys studied my exam papers when they were returned to us, but it was clear I had exceeded their expectations. My exam results ended the unpleasantness, even though they made it clear they considered me an anomaly.

After two years of maths – the boys had five years – I had the highest scores in the national tests, but the head-master insisted that the Prize for Mathematics go to a boy, and so somewhat ironically I was awarded the Needlework Prize instead.

My exam results made it possible for me to change schools at 16, and on my first day I requested the opportu-nity to take pure and applied mathematics. I was told that I could not, even though the boys who transferred with me were allowed to take maths I was not.

Still, my two years of mathematics in my 14th and 15th years were enough some 12 years later I scored a modest 500 in math on the GRE. Subsequently I took advanced courses in statistics for a Masters degree in the Psychology of Reading before transferring to Columbia University for a second Masters and a Doctorate.

It has never been clear to me why I did not allow society to classify me either as working class and therefore of low intelligence, or as a girl, who like all girls, was bad at math-ematics. In the UK both these states-of-being still provoke the upper classes into the bigotry of exclusionary practices that close the doors on girls like me. Rebellion was always to prove them wrong. Whatever the negative expectations of England's pathological class-ridden society, I refused to accept the "lumpenness" ascribed to me.

Years later, when I visited the town in which I had gone to secondary school, a young woman came up to me and said she wanted to thank me. "What for?" I asked, not knowing who she was. "For making it possible for me to take maths," she said. "They let me because you did," she said. "You showed them that girls can do maths."

And so, once again, I say to all you young women who believe you are no good at mathematics, *Master your heart! The very idea that you are bad at math has been inculcated in you through the nefarious practices of the gender that is often called your better half!*

Rebellion is making mathematics your passion. Mathematics, physics, and chemistry are not as difficult as you expect them to be. The curricular practices and requirements in the sciences are determined by men who are also enculturated into not expecting girls and young women to excel at them.

Peter Hampson Ditchfield's 19th century commentary on the positioning of women in the field of mathematics is as relevant in the 21st century as it was when he wrote it.

The book is advertised in *The Antiquary*, November 1886, p.232, by the publisher, Elliot Stock, as a book "of keen insight and intelligence" and "no mean contribution to the political literature". I agree with this accolade and regard the satire as a work that has much to offer participants in courses in political science as well as mathematics, and it contains many hot topics for courses on philosophy *and* education.

There are no spoilers in my introduction to the Reverend Bro Ditchfield's *jeu d'esprit* skit – it is an enjoyable romp, once you know he is taking the mick. More seriously, he has done us a great service 140 years hence by bringing to our attention continued prejudice against women being equal to men in the fields of science and mathematics.

Always remember you are worth more than a thinly disguised licentious part in some rich man's satirical skit. The world will be a better place when there is a 50-50 split.

Denny Taylor

New York
May 26, 2015

Introduction

The lectures, essays, and other matter contained in these pages have been discovered recently in a well-worn desk which was formerly the property of a Lady Professor of Girtham College; and as they contain some original thoughts and investigations, they have been considered worthy of publication.

How they came into the possession of the present writer it is not his intention to disclose; but inasmuch as they seemed to his unscientific mind to contain some important discoveries which might be useful to the world, he determined to investigate thoroughly the contents of the mysterious desk, and make the public acquainted with its profound treasures.

He found some documents which did not refer exactly to the subject of 'Polemical Mathematics;' but knowing the truth of the Hindoo proverb, 'The words of the wise are precious, and never to be disregarded,' and feeling sure that this Lady Professor of Girtham College was entitled to that appellation, he ventured to include them in this volume, and felt confident that in so doing he would be carrying out the intention of the Authoress, had she expressed any wishes on the subject. In fact, as he valued the interests of the State and his own peace of mind, he dared not withhold any particle of that which he conceived would confer a lasting benefit on mankind.

Internal evidence seems to show that the earlier portion of the MS. was written during the period when the authoress was still *in statu pupillari*; but her learning was soon recognised by the Collegiate Authorities, and she was speedily elected to a Professorship. Her lectures were principally devoted to the abstruse subject of Scientific Politics, and are worthy of the attention of all those whose high duty it is to regulate the affairs of the State.

The Editor has been able to gather from the varied contents of the desk some details of the Author's life, which increase the interest which her words excite; and he ventures to hope that the public will appreciate the wisdom which created such a profound impression upon those whose high privilege it was to hear the lectures for the first time in the Hall of Girtham College.

Paper I.

Some Remarks Of A Girtham Girl On Female Education

[This essay upon Female Education was evidently written when the future Professor of Girtham College was still in the lowlier condition of studentship, before she attained that eminence for which her talents so justly entitled her. Its unfinished condition tends to show that it was probably evolved during moments of relaxation from severer studies, without any idea of subsequent publication.]

Oh, why should I be doomed to the degradation of bearing such a foolish appellation! A Girtham Girl! I suppose we have to thank that fiend of invention who is responsible for most of the titular foibles and follies of mankind—artful Alliteration. The two *G*'s, people imagine, run so well together; and it is wonderful that they do not append some other delectable title, such as 'The Gushing Girl of Girtham,' or 'The Glaring Girl of Glittering Girtham.' O Alliteration! Alliteration! what crimes have been wrought in thy name! Little dost thou think of the mischief thou hast done, flooding the world with meaningless titles and absurd phrases. How canst thou talk of 'Lyrics of Loneliness,' 'Soliloquies of Song,' 'Pearls of the Peerage'? Why dost thou stay thine hand? We long for thee to enrich the world with 'Dreams of a Dotard,' the 'Dog Doctor's Daughters,' and other kindred works. Exercise thine art on these works of transcendent merit, but cease to style thy humble, but rebellious, servant a Girtham Girl!

But what's in a name? Let the world's tongue wag. I am a student, a hard-working, book-devouring, never-wearied student, who burns her midnight oil, and drinks the strong bohea, to keep her awake during the long hours of toil, like any Oxford or Cambridge undergraduate. I often wonder whether these mighty warriors in the lists—the class lists, I mean—really work half so hard as we poor unfortunate 'Girls of Girtham.' Now that I am writing in strict confidence, so that not even the walls can hear the scratchings of my pen, or understand the meaning of all this scribbling, I beg to state that I have my serious doubts upon the subject; and when last I attended a soirée of the Anthropological Society, sounds issued forth from the windows of the snug college rooms, which could not be taken as evidences of profound and undisturbed study.

Sometimes I glance at the examination papers set for these hard-working students, in order that they may attain the glorious degree of B.A., and astonish their sisters, cousins, and aunts by the display of these magic letters and all-resplendent hood. And again I say in strict confidence that if this same glorious hood does not adorn the back of each individual son of Alma Mater, he ought to be ashamed of himself, and not to fail to assume a certain less dignified, but expressive, three-lettered qualification. But before those Tripos Papers I bow my head in humble adoration. They sometimes take my breath away even to read the terrible excruciating things, which seem to turn one's brain round and round, and contort the muscles of one's face, and stop the pulsation of one's heart, when one tries to grasp the horrid things.

Here is a fair example of the ingenuity of the hard-hearted examiners, who resemble the inquisitors presiding over the tortures of the rack, and giving the hateful machine just one turn more by way of bestowing a parting benediction on their miserable victims:

'A uniform rod' (it is a marvellous act of mercy that the examiner invented it *uniform*; it is strange that its thickness did not vary in some complicated manner, and become a veritable birch-rod!) 'of length 2c, rests in stable equilibrium' (stable! another act of leniency!), 'with its lower end at the vertex of a cycloid whose plane is vertical' (why not incline it at an angle of 30°?) 'and vertex downwards, and passes through a small, smooth, fixed ring situated in the axis at a distance b from the vertex. Show that if the equilibrium be slightly disturbed, the rod will perform small oscillations with its lower end on the arc of the cycloid in the time

$$4\pi \sqrt{\frac{a\{c^2 + 3(b-c)^2\}}{3g(b^2 - 4ac)}},$$

where 2a is the length of the axis of the cycloid.'

A sweet pretty problem, truly! And there are hundreds of

the same kind—birch-rods for every back! How the examiner must have rejoiced when he invented this diabolical rod, with its equilibrium, its oscillations, its cycloid, and other tormenting accessories. And yet, I suppose, before my days of studentship are over, I shall be called upon to attack some such impregnable fortresses of mathematics, when I hope to be declared equal to some twentieth wrangler, if I escape the misfortune of sharing a portion of the 'wooden spoon.'

Ah, you male sycophants! You would prevent us from competing with you; you would separate yourselves on your island of knowledge, and sink the punt which would bear us over to your privileged shore. Of all the twaddle—forgive me, male sycophants!—that the world has ever heard, I think the greatest is that which you have talked about female education. And the best of it is, you are so anxious about our welfare; you are so afraid that we should injure our health by overmuch mental exertion; you profess to think that our brains are not calculated to stand the strain of continued mental exercise; you think that competition is not good for the female mind; that we are too competitive by nature—too ambitious!

Yes, we are so ambitious that we would enter the lists with those who are asked in Public Examinations to find the simple interest on £1,000 for 5 years at 6¼ per cent.; so ambitious that we would compete with those who are requested to disclose the first aorist middle of $\tau\upsilon\pi\tau\omega$. Oh, think of the mental strain involved in such questions! How it must ruin your health to find out how many times a wheel of radius 6 feet will turn round between York and London, a distance of 200 miles! It is quite wonderful how your brains, my dear male sycophants, can stand such fearful demands upon your intelligence and industry!

But you are so kind to us, so afraid of our health! Really,

we are much obliged to you. If you married one of us, or became our guardian, or left us a legacy, we should then recognise your interest in us, and be very grateful to you for your good advice. But as matters stand, we are quite capable of taking care of ourselves. We will promise not to work too hard, if you will promise not to weary us with your paternal jurisdiction.

But, male sycophants, I want a word with you. Why do you object to our taking degrees, or going in for examinations in order to qualify ourselves for our duties in life? You need not speak out loud if you would rather not. Are you not just a little afraid that we might eclipse you? And it is not pleasant to be beaten by a woman, is it? And then you profess to think that we ought to be all housewives and cooks, and knitters of stockings, and sewers-on of our husbands' buttons; but what if we have no husbands, no buttons to sew? And is it not a little selfish, my dear male sycophant, to wish to keep us all to yourself? to attend upon the wants of the lords of creation, who often distinguish themselves so much in the domain of science?

Now, look me straight in the face (no shirking, sir!). Is it not jealousy—green-eyed, false-tongued jealousy—which saps your generous instincts, and makes you talk rubbish and nonsense about strains, and brains, and ambition, and the like? And if that is not hypocritical, I do not know what is.

Well, good-day to you, male sycophant! I really have not time to indulge myself in scolding you any more. You are a good creature, no doubt; and when you have shown us what you can do, and can estimate the capacity of the female brain, and take a common-sense view of things, we will recognise your privilege to speak; and when I am the presiding genius of Girtham College, I will grant you the use of our hall for the purpose of lecturing to us on 'Women's Rights,' or, as you may prefer to entitle your discourse, 'Men's Wrongs.'

Oh, this is shameful! I really am very sorry. Here have I been wasting a good half-hour in dreaming, and slaying an imaginary enemy with envenomed words and frequent dabs of ink. If I cannot concentrate my mind more on these mathematical researches, I fear a dreadful 'plough' will harrow my feelings at the end of my sojourn in these halls of learning.

Concentration! How many of our words and ideas and thoughts are derived from that primal fount of all arts and sciences—mathematics! Here is one which owes its origin to the mathematically trained mind of some early philological professor, who had learnt to apply his scientific knowledge to the enrichment of his native tongue. He quoted to himself the words of the Roman poet:

> 'Ego cur, acquirere pauca
> Si possum, invideor, cum lingua Catonis et Ennî
> Sermonem patrium ditaverit, et nova rerum
> Nomina protulerit? Licuit, semperque licebit.'

His mind conceived endless figures of circles and ellipses scattered promiscuously over the page, defying the attempts of the student to reduce them to order. What must he do before he can apply his formulæ and equations, determine their areas, or describe their eccentric motion? He must reduce them to a common centre, and then he can proceed to calculate the abstruse problems in connection with the figures described. They may be the complex motions of double-star orbits, or the results of the impact of various projectiles on the tranquil surface of a pool. It matters not—the principle is the same; he must concentrate, and reduce to a common centre.

This is the great defect of those who have no accurate mathematical knowledge; they cannot concentrate their minds with the same degree of intensity upon the work which lies before them. Their thoughts fly off at a tangent, as mine do very often; but then I have not been classed yet in the Tripos; and, O male poetical sycophant, you may be right after all when you say:

> 'O woman! in our hours of ease
> Uncertain, coy and hard to please,
> As variable as the noon-day shade.'

Yes, as variable as the most variable quantities x, y, z. I, a student of Girtham College, blush to own that my thoughts very often fly off at a tangent.

'Fly off at a tangent!' All hail to thee, most noble mathematical phrase! Here is another fine mathematical expression, plainly exemplifying the action of centrifugal force. The faster the wheel turns, the greater is the velocity of the discarded particles which fly off along the line, perpendicular to the radius of the circle. The world travels very fast now; the increased velocity of the transit of earthly bodies, the rate at which they live, the multiplicity of engagements, etc., have made the social world revolve so fast that the speed would have startled the torpid life of the last century. And what is the result? Men's thoughts fly off at a tangent; they are unable to concentrate their minds on any given subject; they are content with hasty generalisms, with short magazine articles on important subjects, which really require large volumes and patient study to elucidate them fully.

What we want to do is to increase the attractive force, in order to prevent this tangential motion—to increase the *force of gravity.*

'Well,' says the young lady who loves to revel in the 'Ghastly Secret of the Moated Dungeon,' or the 'Mysteries of Footlight Fancy,' 'you are *grave* enough. Pray don't increase your gravity!'

Thank you, gentle critic. I will, in turn, ask you one favour. Leave for once the 'Mysteries of Footlight Fancy;' seek to know no more 'ghastly secrets,' and increase *your gravity*—your mental weight; and hence your attraction in the eyes of all who are worth attracting will be marvellously increased, by understanding a little about Newton's law of universal gravitation, and don't fly off at a tangent.

[Editorial Note] At the end of this portion of the MS. the editor of these papers discovered a photograph which, from subsequent inquiry, proved to be that of the accomplished authoress of the above reflections. The face is one of considerable beauty, with eyes as clear, steadfast, and open as the day. There is a degree of firmness about the mouth, but it is a sweet and pretty one notwithstanding; and a smile, half scornful, half playful, can be detected lurking about the corners of the lips, which do not seem altogether fitted for pronouncing hard mathematical terms and abstruse scientific problems

This photograph might have been the identical one which nearly brought an enamoured youth into grave difficulties by its secretion in the folds of his blotting-

paper during examination. The said enamoured youth had evidently placed it there for the sake of its inspiring qualities; and it was said that all his hopes of gaining the hand of the fair original depended upon his passing that same examination. But the wakeful eye of a stern examiner had watched him as he turned again and again to consult the sweet face which beamed from beneath his blotting-paper; and he narrowly escaped expulsion from the Senate-house on the charge of 'cribbing.' Certainly he took a mean advantage of his fellow-sufferers, if this were the identical photograph, for it portrays a most inspiring face. Forgive us, lenient reader; one moment! There—thank you—we have done. And now we will proceed to disclose the researches and original problems which the MS. contains.

Evidently the collegiate authorities were not slow in recognising the talents of the assiduous student, and elected her without much delay to a Professorship of Girtham. In this capacity the learned lady delivered several lectures, of which the second MS. contains the first of the series.

Paper II.

Lecture On The Theory Of Brain Waves And The Transmigration And Potentiality Of Mental Forces

Professors and Students of the University of Girtham, my Lords, Ladies, and Gentlemen,—I have the honour to bring before you this evening some original conceptions and discoveries which have been formulated by me during my researches in the boundless field of mathematical knowledge; and though you may be inclined at first to pronounce them as somewhat hastily conceived hypotheses, I hope to be able to demonstrate the actual truth of the propositions which I shall now endeavour to enunciate.

It is with some feelings of diffidence that I stand before so august an assembly as the present; and if I were not actually convinced of the accuracy of my calculations, I should never have presumed to appear before you in the character of a lecturer. But '*Magna est veritas, et prævalebit.*' I cast aside maiden timidity; I clothe myself in the professorial robe which you have bestowed upon me, and sacrifice my own feelings on the altar of Truth.

I have been engaged, as you are doubtless aware, for some years in the pursuit of mathematical research, exploring the mines of science, which have of late been worked very persistently, but often, like the black diamond mines, at a loss. Concurrently with these researches, I have speculated on the great social problems which perplex the minds of men, both individually and collectively. And I have come to the conclusion that the same laws hold good in both spheres of work; that methods of mathematical procedure are applicable to the grand social problems of the day and to the regulation of the mutual relations which exist between man and man.

Take, for example, the Force of public opinion. Of what is it composed? It is the Resultant of all the forces which act upon that which is generally designated the 'Social System.' Public opinion is a compromise between the many elements which make up human society; and compromise is a purely mechanical affair, based on the principle of the Parallelogram of Forces. Sometimes disturbing forces exert their influence upon the action of Public Opinion, causing the system to swerve from its original course, and precipitating society into a course of conduct inconsistent with its former behaviour; and it is the duty of the Governing Body to eliminate as far as possible such disturbing forces, in order that society may pursue the even tenor of its way.

Professors, we have one great problem to solve; and all questions social, political, scientific, or otherwise, are only fragments of that great problem. All truths are but different aspects of different applications of one and the same truth; and although they may appear opposed, they are not really so; and resemble lines which run in various directions, but lovingly meet in one centre.

Now, let us take for our consideration the secret influence which men exert upon each other, apart from that produced by the power of speech (although that would come under the same general law).

As mathematicians, you are aware that the undulatory theory of light and heat and sound are now accepted by scientific men as the only sure basis of accurate calculation. We know that the rays of light travel in waves, and the equation representing the waves is

$$y = \frac{a}{r} \sin \frac{2\pi}{\lambda} (vt - r),$$

where y is the disturbance of the ether, a the initial amplitude, r the distance from the starting-point, λ the wavelength, and v the velocity of light. Sound and heat likewise have much the same form of equation.

Now, I maintain that the waves of thought are governed by the same laws, and can be determined by an equation of the same form. You are aware that in all these equations a certain quantity denoted by λ appears, and varies for the different media through which the sound, or light, or heat passes, and which must be determined by experiment. Now, in my equation for brain waves, the same quantity λ appears which must be determined by the same method—by *experiment*.

But how is this to be done? After mature deliberation and much careful thought, I have discovered the method for finding λ. This method is *mesmerism*. We find the ratio of brain to brain—the relative strength which one bears to another; and then by an application of our formula we can actually determine the wave of thought, and read the minds of our fellow-creatures.

An unbounded field for reflection and speculation is here suggested. Like all great discoveries, the elements of the problem have unconsciously been utilized by many who are unable to account for their method of procedure. For example, thought-readers, mesmerists, and the like, have unconsciously been working on this principle, although lack of mathematical training has prevented them from fully mastering the details of the problem. Hence in popular minds a kind of mystery has hung about the actions of such people, and excited the curiosity of mankind.

The development of this theory of brain waves may be of great practical utility to the world. It shows that great care ought to be exercised in the domain of thought, as well as that of speech. For example: A man has made a startling discovery, from which he expects to receive considerable worldly advantage. He would be careful not to disclose his discovery in speech to his acquaintances until his plans are sufficiently matured, lest they should impart it to the world, patent his device, and reap the reward. But while he is endeavouring to talk carelessly about it, the wave of thought may be travelling from brain to brain, suggesting the existence of the discovery; and if the conditions are favourable, and λ sufficiently small, it is possible that the idea itself may be conveyed. Of course the more complicated the discovery, the less likely would the wave convey the conception.

Or suppose that one of the learned professorial body of our sister university should conceive an attachment for a lady-student of Girtham College (of course a very improbable supposition!), and the infatuated *savant* became somewhat jealous of another learned lecturer of the same college (another improbability!), the fact of his jealousy would be imparted to the latter by a wave of thought, and might cause considerable confusion in the serene course of love or science.

The fact of the existence of the wave is indisputable. What do all the stories of impressions and double-sight teach us? How could the intelligence of the death of Professor Steele have been conveyed to his friend and fellow-student, Professor Tait—the one at Cambridge, the other at Edinburgh—were it not for the existence of some wave, which, like that of electricity, wings its rapid flight unobserved by human eyes? Are all the records of the Psychical Society only myths and legends bred of superstitious fancy? It were hard to suppose so.

But if, gentlemen, and ladies especially, you wish to keep your secret discoveries to yourselves, watch over your thoughts as well as your words; for my researches prove, and the universal experience of mankind corroborates the fact, that some portion of your inmost thoughts and secret desires are understood by your neighbours (especially when λ is small!); that they travel along the waves which I have attempted to indicate; and if you would desire to extend your influence in the world, probe the secret instincts of mankind, and prevent yourself from being deceived and wronged—study the art and science of Brain Waves.

[Editorial Note] The following verses of rather doubtful merit were found in connection with the previous MS. They were evidently written by a different hand; but inasmuch as they were deemed worthy of preservation by the learned owner of the sealed desk, we venture to publish them. They are closely connected with the previous lecture, and were evidently composed by an

admirer of the fair lecturer who did not share her love for scientific research.

Wavelet,[1] wing thy airy flight;
Let thine amplitude be great;
Tell her all my thoughts to-night,
How I long to know my fate.

All the fields of Mathematics
I have roamed at her decree;
From Binomial and Quadratics,
To the strange hyperbole.[2]

I have soared through Differential,
Deeply drunk of Finite Boole[3]
Though its breath is pestilential,
Reeking of the hateful School.

I have tried to shape a Conic,
Vainly read the Calculus;
But my feebleness is chronic,
Morbus Mathematicus.

All my curves are cardioidal;
I confuse my x and ys,
Which they say is suicidal;
And my tutor vainly sighs.

Wavelet, tell her how I love her,
As she mounts her learned throne;
And that love I hope may cover
All the failings which I own.

P. Hampson

Wavelet, cry to her for pity;
Bid her end this bitter woe;
I might do something 'in the city,'
But never pass my Little-go.

Notes

1. We presume this is addressed to an imaginary brain wave.
2. We observe here the dash of an indignant pen, and *a* substituted for *e*. But now the rhyme is spoiled. Gentle Muse, thou art sacrificed by the stern hand of Mathematical Truth!
3. Query: Does the writer refer to the learned treatise on Finite Differences by Professor Boole?

Paper III.

Lecture On The Social Properties Of A Conic Section, And The Theory Of Polemical Mathematics

Most Learned Professors and Students of this University,—From the interest manifested in my first lecture, I conclude that my method of investigation has not proved altogether unsatisfactory to you, and I hope ere long to produce certain investigations which will probably startle you, and revolutionize the current thought of the age.

The application of mathematics to the study of Social Science and Political Government has curiously enough escaped the attention of those who ought to be most conversant with these matters. I shall endeavour to prove in the present lecture that the relations between individuals and the Government are similar to those which mathematical knowledge would lead us to postulate, and to explain on scientific principles the various convulsions which sometimes agitate the social and political world.

Indeed, by this method we shall be able to prophesy the

future of states and nations, having given certain functions and peculiarities appertaining to them, just as easily as we can foretell the exact day and hour of an eclipse of the moon or sun. In order to do this, we must first determine the *social properties of a conic section*.

For the benefit of the unlearned and ignorant, I will first state that a cone is a solid figure described by the revolution of a right-angled triangle about one of the sides containing the right angle, which remains fixed. The fixed side is called the axis of the cone. Conic sections are obtained by cutting the cone by planes. It may easily be proved that if the angle between the cutting plane and the axis be equal to the angle between the axis and the revolving side of the triangle which generates the cone, the section described on the surface of the cone is a parabola; if the former angle be greater than the latter, the curve will be an ellipse; and if less, the section will be a hyperbola.

But the simplest conic section is, of course, a circle, which is formed by a plane at right angles to the axis of the cone; and the simplest circle is that formed by a plane passing through the apex of the cone. All this is simple mathematics; and let beginners consult more elementary treatises than this one to satisfy themselves on these points. But if they will assume these things to be true, they will know quite enough for our present purpose.

The simplest conic section of all has been proved to be a *point*. Now, this represents the simplest and original form of society, a *single family*. 'It is not good for man to be alone' was the first observation made by the wise Creator upon the rational creature whom He had introduced into Paradise as its lord. Marriage is the rudiment of all social life, from which all others spring, out of which all others are developed. Around the parents' knees soon cluster a group of children, and in their relation to each other we discern the earliest forms of law and discipline—the bonds by which society is held together. When the children grow up, separate households are formed; and then the multiplication of families, the congregating of men together for purposes of security and mutual advantages in division of labour; and thus is gradually formed a state, which is only the development of the family—the king representing the parent, and ruling on the same principle.

Mathematically speaking, our plane no longer passes through the apex. The point represented the single family; but keeping the plane horizontal, we move it along the axis, the sections will become *circles*, which represent mathematically the next simplest form of society, where the centre is the seat of government, which is connected with each individual member of the social circle by equal radii. The social property of a circle is that of a monarchical government in its purest and simplest form. The larger the circle becomes (*i.e.*, the further you move the plane from the apex), the greater the distance between the individual and the monarch. Therefore, the more independent the monarchy becomes, and the less influence do individuals possess over the ruling power.

Hence, we may infer that as years roll on, the government will become more despotic; but the stability of the country diminished, and probably some individual particle, when sufficiently withdrawn from the attraction of the central head, will begin to revolve on its own account, and spontaneously generate a government of its own.

We may, therefore, conclude from mathematical reasoning that an unlimited monarchy, though advantageous for small states, is not a safe form of government for a large or populous country, inasmuch as the people do not derive much benefit from the sovereign; the mutual attraction, which ought to exist in a flourishing state between the ruler and the ruled, is weakened; and the isolation of the monarch tends to make him still more despotic. As a practical example of the truth of the foregoing statement, I may mention the present condition of Russia, which shows that the result of an unlimited monarchy, in a large and unwieldy social circle, is such as we should have reasonably expected from mathematical investigations.

Invariably, under the circumstances which I have described, the country will become disorganized; the sovereign will cease to have any power over the people, and the country will become a chaos, without order, influence, or power.

When the centre of a conic section moves along the axis of the curve to infinity, banished by the mutual consent of the individual particles which compose the curve, or the nation, a figure is formed, called a *parabola*. This is the curve which the most erratic bodies in the universe describe in space, as they rush along at a speed inconceivable to human minds, and are supposed to produce all kinds of mischief and injury to the worlds whose courses they wend their way among.

This curve, then, represents the position which the nation assumes when the constituted monarchy, the centre of the system, has been *banished to infinity*. A revolution has occurred; the monarch has been dethroned; and it is not hard to see that the same erratic course which the comet pursues in its flight, is observable with respect to the social system which is represented by a parabola.

We observe with eager scrutiny the wanderings of these erratic comets. They appear suddenly with their vapoury tails; sometimes they shine upon us with their soft, silvery light, brilliant as another moon; sometimes they stand afar off in the distant skies, and deign not to approach our steady-going earth, which pursues its regular course day by day, and year by year. Then, after a few days' coy inspection of our planet from different points of view, they fly to other remote parts of the universe, and do not condescend to show themselves again for a hundred years or so. Such is the erratic conduct of a heavenly body whose course is regulated by a parabolic curve.

We may look for similar eccentric behaviour on the part of a community, nation, or state, whose centre is at infinity, whose constitution has been violently disturbed, and whose monarchy is situated in the far-off regions of unlimited space. The erratic course of Republican rule is proverbial. There is no stability, no regularity. To-day we may observe its brilliancy, which seems to laugh at and eclipse the sombre shining of more steady and enduring worlds; but ere to-morrow's moon has risen, it may have vanished into the regions of eternal night, and we look for its bright shining light in the councils of the nations, but it has ceased to shed its rays, and we are disappointed.

Sometimes it is asked, with fear and trembling: 'What would be the effect if our earth were to come in contact with the tail of a comet? Should we be destroyed by the collision, and our ponderous world cease to be?' But we are assured that no such disastrous results would follow. We have already passed through the tails of many comets, but we have not discovered any inconvenient change in our ordinary mode of procedure. It is probable that the comet's tail is composed of no solid substance.

We may therefore infer by analogy that a Republican State would not offer any powerful resistance if it were to come into collision with a nation possessing a more settled form of government. A shower of meteoric stones, like passing fireworks, might take place; but beyond that nothing would occur to excite the fear, or arouse the energies of the more favoured nation.

As an example of the weakness of a Republican State I may mention France. There we see an industrious race of people, endowed with many natural gifts and graces, a country rich and productive; and yet, owing to the unsettled nature of its government, all these natural advantages are neutralized; its course amongst the nations is erratic in the extreme, a spectacle of feeble administration; and it would offer no more resistance to a colliding Power than the empty vacuum of a comet's tail. This example will demonstrate to you the truth of our theory with regard to the instability of a social system which is geometrically represented by a parabolic curve.

We will now turn from this picture of insecurity and unrest to another figure which possesses most advantageous social properties. I refer to the ellipse. An ellipse is a curve formed by the section of a cone by a plane surface inclined at an angle to the vertical axis of the cone, greater than the angle between the axis and the generating line.

Now, this is a curve which possesses most attractive properties. It is the curve which the earth and other planetary orbs describe around the centre of the solar system, as if nature intended that we should take this figure as a guide in choosing the most advantageous social system. It possesses a centre, C, in view of all the particles which compose the curve, and connected with them by close ties. It has two foci, S and S', fixed points, by the aid of which we may trace the curve.

In the interpretation of this figure, the centre of the curve represents the throne of monarchy. There is no tendency here to revolutionize the State, to banish the ruling power, and institute a Republican form of government; but inasmuch as we saw the weakness of an absolute monarchy in large and populous States, as represented by the circle, the wisdom of an elliptical social system has ordained that there shall be two foci, or houses of representatives of the people, who shall assist in regulating the progress of the nation. Here we have a limited monarchy; the throne is supported by the representatives of the people; and the nearer these foci of the nation are to the centre (*i.e.*, in mathematical language, the less the *eccentricity* of the curve), the more perfect the system becomes—the greater the happiness of the community.

In cases where the *eccentricity* becomes very great, the beauty of the curve is destroyed, and ultimately the ellipse is merged into one straight line. Most learned Professors, here we have a terrible warning of the awful result of too much eccentricity. Whether we regard the life of the nation or of the individual, let all bear in mind this alarming fact, that eccentricity of thought, habit, or behaviour may result, as in the case of this unfortunate ellipse, which once presented such fair and promising proportions to the student's admiring gaze, in the 'sinister effacement of a man,' or the gradual absorption of a State into an uninteresting thing 'which lies evenly between its extreme points.'

The great examples of Bacon, of Milton, of Newton, of Locke, and of others, happen to be directly opposed to the popular inference that eccentricity and thoughtlessness of conduct are the necessary accompaniments of talent, and the sure indications of genius. I am indebted to Lacon for that reflection. You may point to Byron, or Savage, or Rousseau, and say, 'Were not these eccentric people talented?' 'Certainly,' I answer; 'but would they not have been better and greater men if they had been less eccentric—if they had restrained their caprice, and controlled their passions?' Do not imagine, my young students of this university, that by being eccentric you will therefore become great men and women of genius. The world will not give you credit for being brilliant because you affect the extravagances which sometimes accompany genius.

Some of you ladies, I perceive, have adopted a peculiar form of dress, half male, half female; or, to be more correct, three-fourths male, and one-fourth female. Do not imagine that you will thus attain to the highest honours in this university by your eccentricity, unless your talents are hid beneath your short-cut hair, and brains are working hard under your college head-gear. As well might we expect to find that all females who wear sage-green and extravagant æsthetic costumes are really born artists and future Royal Academicians. It is apparent that many aspirers to fame and talent are eager to exhibit their eccentricities to the gaze of the world, in order that they may persuade the multitude that they possess the genius of which eccentricity is falsely supposed to be the outward sign.

I may remark in passing that the eccentricity of a parabolic curve is always *unity*. What does this prove? You will remember that a Republican State is represented by a parabola. Therefore, however such a nation may strive to alter its condition, and secure a settled form of government, its eccentricity will always remain the same. It will always be erratic, peculiar, unsettled; and this conclusion substantiates our previous proposition with regard to the condition of a social system represented by a parabola.

With regard to other advantages afforded by an elliptical social system, we will defer the consideration of this important subject until my next lecture.

Paper IV.

The Social Properties Of A Conic Section, And The Theory Of Polemical Mathematics— *[continued]*

Most learned Professors and Students of this University,—You have already gathered from my preceding lecture my method of procedure in the investigation of the corresponding properties of curves and States. You have perceived that we have here the elements of a new science, which may be extended indefinitely, and applied to the various departments of self-government and State control.

This new science of polemical mathematics is in itself an extension of the *principle of continuity*, for the discovery of which Poncelet is so justly renowned. We can prove by geometry that the properties of one figure may be derived from those of another which corresponds to it; and the new science teaches us that if we can represent, by projection or otherwise, a society of particles or individuals on a plane surface, the properties of the State so represented are analogous to the properties of the curve with which it corresponds.

It is only possible for me to touch upon the elements of the science in these lectures, but I hope to arouse an interest in these somewhat unusual complications and curious problems, that you may hereafter make further discoveries in this unexplored region of knowledge, and that the world may reap the benefit of your labours and abstruse studies. I have already, in my previous lecture, touched upon the social properties of the parabola, and examined the constitution of erratic curves and eccentric nations. It is my intention to-day to speak of similar problems which arise with reference to elliptical States.

But, first, let me answer an objection which may have occurred to your minds. Am I wrong in my calculations in attributing too much to the power and usefulness of forms of government? Does the well-being and happiness of a nation depend on the government, or upon the individuals who compose the nation? Most assuredly, I assert, they rest upon the former. Men love their country when the good of every particular man is comprehended in the public prosperity; they undertake hazard and labour for the government when it is justly administered.

When the welfare of every citizen is the care of the ruling power, men do not spare their persons or their purses for the sake of their country and the support of their sovereign. But where selfish aims are manifest in Court or Parliament, the people care not for State officials who are indifferent to their country's weal; they become selfish too; Liberty hides her head, and shakes off the dust of her feet ere she leaves that doomed land, and the stability, welfare, and prosperity of that country cease.

I might refer you to many a stained page of national history in order to prove this. Compare the closing chapters of the life of the Roman Empire with the record of the brave deeds of its ancient warriors and valorous statesmen. Grecian preeminence and virtue died when liberty expired. I agree with Sidney when he writes that it is absurd to impute this to the change of times; for time changes nothing, and nothing was changed in those times but the government, and that changed all things. These are his words: 'As a man begets a man, and a beast a beast, that society of men which constitutes a government upon the foundation of justice, virtue, and the common good, will always have men to promote those ends; and that which intends the advancement of one man's desires and vanity will abound in those that will foment them.' I may not, therefore, be altogether wrong in attributing the prosperity and well-being of a nation to the form of government which it possesses.

We will now proceed to the consideration of the social advantages which an elliptical State affords. This is the form of government and social position which we, as a nation, at present enjoy; and from mathematical considerations I am of opinion that it is the best, and hope that no change will ever be made in our constitution. You may remember that I have previously stated that an ellipse has a centre and two foci, in view of all the particles which compose the curve, and connected with them by close ties. The centre, in the projected figure, represents the monarchy, which is limited; and the government is carried on by the aid of the two houses of representatives of the people, depicted in the projection by the two foci.

Now the social advantages of the ellipse are given by the fact that the sum of the distances of any point from the foci is always constant. No particle is left out in the cold; no one does not possess the advantages of a social government. Though his distance may be far from the Upper House, he has the advantage of nearness to the Lower, and *vice versâ*. The sum of the distances is constant.

The extinction of one focus, the House of Lords, for example, would create a complete disorganization of the whole system: the other focus would set up a powerful magnetic attraction, and a curious bulb-shaped curve would be evolved, very different from the beautiful symmetrical form which the original figure presented to the eye. The centre of the system would be disturbed; and it is probable that ere long it would disappear along the axis and be vanished to infinity. Thus the curve would become a parabola.

This is the alarming result of the extinction of one focus. Abolish the House of Lords, and you will soon find that the Throne will be disturbed; the State will become disorganized; the nation will become confused by the magnetic force of the Lower House, uncounteracted by any other attraction; and very soon a complete revolution of the whole system will set in: the monarch will be dethroned, and a Republican form of government, with all the eccentricities of a parabolic course, will take the place of a more orderly and settled constitution.

This is a plain deduction from our mathematical investigations; and it behoves all our statesmen, our philosophers and great men, our fellow-citizens and the humblest artisans in our manufacturing towns, to weigh well this alarming result of the abolition of that House which has been threatened with destruction; and to ascertain for themselves the truths upon which my proposition and reasoning rest.

I have already observed that the fact that the earth's

orbit and that of other planets are in the form of ellipses; that the curvature of the earth is nearly the same, ought to guide us in choosing this particular curve as a model of the projection of a complete and most advantageous social system.

The circle described on the major axis of an ellipse, is called the *auxiliary circle*, and affords much assistance in the investigation of the properties of an ellipse. As we have already shown, the circle represents the simplest form of monarchical government. Hence, if we compare the form of government represented by an ellipse (*i.e.*, such as we now enjoy) with that of a system where the king is the only governing power, we may obtain great assistance in solving complicated political problems.

In all conics there is a straight line called the 'directrix,' which represents in social or polemical science the laws of the nation, and plays a prominent part in the mutual relations of the individual particles. For instance, in the case of the parabola, the distance of any particle from the directrix is equal to its distance from the focus.

From this we may conclude that if an individual deviates at all from the path which the laws (or, directrix) indicate, if he does not show true respect to the decrees of the focal government, and preserve the true position between them, directly he is found deviating from his course, he is quickly banished to a less enlightened sphere. In an ellipse there is less likelihood of his straying away from the course which the directrix points out, on account of the two-fold guidance which he receives from the two foci.

The following curious problem may be noticed. If a parabola roll on another parabola, their vertices coinciding, the focus of the first traces out the directrix of the second.

Here we come to the consideration of the international

relationship of States. Two nations have the same form of government (in this example this form is Republican); their policies coincide: we may conclude from this proposition that the course which the government of one nation will pursue, will be that which is prescribed by the laws of the other.

The subject of the contact of curves presents many interesting problems with reference to Polemical Science, and may be extended indefinitely. It is well known that there are different orders of contact, which are designated as the *first*, *second*, or *third* order. This last order may be termed the 'marriage of curves,' cemented by the osculating circle, or 'wedding-ring;' and when two nations have contact of the third order, they have formed a very close alliance, and by calculation we can obtain the *radius of curvature*, or size of the wedding-ring, by means of which they may be united.

The theory and nature of contact constitute a branch of our newly discovered science which we commend to the careful consideration of those who have undertaken the difficult and perplexing study of international law. Alas! too many States refuse this friendly contact, and, consequently, *cut* each other, instead of blending in sweet accord. Their peace is at best an armed neutrality; and if they have contact of only the *first* or *second* order, we can prove mathematically that they are sure to intersect in some other point or points; and divergence of policy and disturbed relations are the results. Contact of the *third*, or *highest*, *order* is the only safe position for two allied, or contiguous, States.

With your permission I will add a few words to those I have already uttered with regard to the directrix. As necessary as the directrix is to the curve, so are the corresponding laws to the State. I will prove this fact by a few examples. English people have laws, and know how to obey them;

therefore their numbers increase; they thrive and are prosperous. A friendly critic of another nation has said that the reason why Englishmen rule the world, is because they know how to obey. On the other hand, the gipsies have no laws; hence they become fewer and less powerful. What is the condition of all tribes and nations which are not governed by laws? They invariably remain poor and miserable. They are in want of a directrix; and if we could supplement the gift with foci and centre, they would soon emerge from their savage condition, and become more civilized.

I have omitted to mention the hyperbolic form of government. The curve formed by the intersection of the surface of a cone with a plane will be a hyperbola, when the inclination of the cutting plane to the axis of the cone is less than the constant angle which the generating line forms with the axis. It is manifest that the plane will thus intersect the higher cone, and produce the figure which is known to mathematicians as the hyperbola.

We may hence deduce the following property of the corresponding hyperbolic State. We take cognizance of that higher cone with which the mundane affairs of the lower cone are closely connected. As an example of this system we may mention the vast temporal rule and power of the Papal Throne, which formerly exercised such marvellous sway over the nations of Europe. By an appeal to a Higher Authority than that of earthly kings and potentates was this rule exercised; but its hyperbolic form is fast passing away, and degenerating into that of a circle with indefinitely small radius. We shall not, therefore, discuss the complex polemical problems which a hyperbolic State suggests.

I will now mention a few problems which are easily capable of proof, and deduce from them the necessary conclusions which must follow when we apply our newly discovered principles of polemical science.

1. 'If from any point in a straight line a pair of tangents be drawn to an ellipse, the chords of contact will pass through a fixed point.'

I will not trouble you with the proof of this proposition, as it is evident to all mathematicians, and can easily be demonstrated. But mark well the deductions, when we interpret this mathematical language in correct polemical terms.

A State, through various convulsions of its own, has merged into a condition represented by a straight line, having lost its symmetry, its beauty, its curvilinear proportion. An individual unhappily situated in this unfortunate community regards with longing eyes the prosperous condition of those who enjoy the social advantages of a settled form of government, and other blessings which accompany elliptical jurisdiction and laws. [Two tangents are drawn to an ellipse.] No matter where the individual may be in the unhappy envious straight line, the result of his reflection will be the same. Sympathetic chords are drawn, joining the points of contact of the tangents with the curve; they all pass through a fixed point.

All these conclusions of the various individuals on the straight line will be the same. All are of opinion that the elliptical form is the best; and they mourn in secret over the sad events which have occurred in their own national life, their eccentricity, their lawlessness, when they see the advantages which their more staid and sober-minded neighbours so freely enjoy.

2. The normal at any point of an ellipse bisects the angle between the focal distances of that point.

The normal is the perpendicular from the point on the

major axis; it is the line of thought directed by the observance of just laws and rules. Hence this proposition shows that the individual citizen, when guided by sound judgment, regards with equal favour and entire approval the existence of both foci, or Houses of Legislature. He considers that both are necessary to his comfort, and the right regulation of the State's welfare. He cares not for the *abnormal* condition of those who talk as if the existence of either House were unnecessary to his country's weal, and bestows a pitying glance on those wandering lights, or disturbed erratic governments, which do not possess the advantages which from experience he has learned to love and to respect.

No matter what his condition may be, the same opinions are held by all classes, all ranks and degrees; and if a self-opinionated particle think otherwise, he ought to be transferred to a less enlightened sphere, and migrate to a parabolic state, or uninteresting straight line. And when he has changed his location, he will look back on his old home and old surroundings with longing eyes and an aching heart, thinking of the blessings he has lost by his own rash act. This can be proved mathematically. He looks for an ideal state of society, leaps after the shadow his fancy has depicted; and when he finds himself outside his former state, he looks back with longing eyes at the once-scorned focus.

What is the focus of a perpendicular on the tangent of an ellipse from any external point? Can it not be proved to be a *circle*? That is to say, he will be more conservative than ever. He would like to return to a primitive form of government. Farewell to his wild schemes and revolutionary measures! Farewell to his disestablishments, abolitions, and suppressions! The throne and government have new attractions in his eyes; loyalty, a new feeling, asserts its benign influence; and if he could return to his former position, his normal conduct would be straighter than ever, for by sad experience he has learned the value of those things which he once despised.

But we need not depend upon one proof alone. Exactly the same result may be obtained from the well-known proposition which states that 'the angle between the tangent from any external point and the focal distance is equal to the angle between the other tangent and the focal distance.'

3. The same opinions are often held by individuals in quite different walks and classes of life. Let these individuals be represented by points on an ellipse. Join these, and we have a system of parallel chords. Draw a straight line through the middle points of these chords, and lo! it will always pass through the centre. This shows that the central thought of all people is directed to the sovereign—that *loyalty* is inherent in the hearts of those who recognise elliptical laws.

I will conclude this lecture with a few remarks on the nature and properties of the *radical axis*. This name was first given, I believe, by M. Gaultier, of Tours, and for a full account of its nature I refer you to the *Journal de l'École Polytechnique*, xvi., 1813. The radical axis of two circles is the line perpendicular to the line joining the centres, from any point of which the tangents to the circles are equal. Let us suppose that one circle becomes a point, and that this point is situated on the circumference of the first circle. What is the result? The radical axis becomes the tangent to the circle. Hence we may conclude that in a social system of monarchical government the radical axis is perpendicular to the line attaching the individual with the monarch.

Therefore we may conclude that the radical axis indicates a tendency of particles, or individuals, to fly off at a tangent, at right angles to the connecting-link between the individual and the king. When any motion takes place, this is evident, and this tendency is called centrifugal force. Sad is it for the State when this force is called into play, and the radical axis is a standing menace to the stability of States and nations. The only way to counteract its baneful, disturbing influence is to increase the attraction of the monarch on the individual, which nullifies the former force, and prevents further mischief.

This is the method which nature itself adopts in the motions of the planetary worlds; the attraction of the sun prevents any disturbance which might be caused in the course of the planets by the action of centrifugal force, and nature suggests this plan for our adoption. Increase the attraction of the Throne; rigidly connect each individual by the strong chords of affection, advantage and utility with the ruling power; and then, though the radical axis may be there, it will cease to indicate any motion along it, it will not prevail over the counteracting influence of loyalty, and the stability of the social system and the happiness of the individuals will be the results.

> 'I would serve my King,
> Serve him with all my fortune here at home,
> And serve him with my person in the wars;
> Watch for him, fight for him, bleed for him, die for him,
> As every true-born subject ought.'

This, most noble professors, is the language of true patriotic loyalty. Let the monarch be loved and loving, let the laws be just and equal, happy will be the people, prosperous the realm. There are those who counsel different things, and preach sedition and the breaking-up of laws; but those who advocate such doctrines lack that judicial mathematical training which we, students and professors of Girtham College, have acquired.

If polemical mathematics, the science of the future, should become more widely studied; if its results were disseminated far and wide; above all, if the proper position which women ought to occupy in the counsels of the nation were assigned to them, we should hear less of these wild schemes and foolish theories, and the influence of women would tend greatly to promote the stability and security of the State.

Why, let me ask, should woman be excluded from that position which is so justly hers? from those duties which she can discharge so faithfully? It has been said that if we wish to know the political and moral condition of a State, we must ask what rank women hold in it. We are told that women have more strength in their looks than men have in their laws. Why, then, do men debar her from those fields of occupation wherein she may labour for the nation's good, and use her influence, which they acknowledge to be great, in those callings wherein she may most easily benefit the State, and the country she so ardently loves?

At some future time I hope to speak more fully on this subject; and in concluding this lecture, I will remark that English politics need a leavening influence which will counteract the evil tendencies and corrupt theories which, in spite of our advantageous social system, at present exist; and this leavening influence will be best produced by the admission of those into the counsels of the nation who are acknowledged to have a benign and healthy influence—the women of England. Let women have their proper share in the government of the country, and I have no fear lest we shall preserve our elliptical constitution, and all the advantages which we at present enjoy.

[Editorial Note]—In the bundle of papers which contained the foregoing lectures, some letters of great interest were found, which show that the fame of the learned Lady Professor of Girtham College had already gone abroad, and attracted the attention of the leading statesmen of the day. It is to be regretted that the answers to these letters are not forthcoming, as it might be proved from them that the science of polemical mathematics has already influenced the minds of our legislators in their conduct of affairs at home and abroad. The following letter is of unique interest, and may be taken as evidence of the favourable impression which this new science has made on the mind of one of our greatest thinkers and statesmen:

Downing Street,
May, 18—

My dear Lady Professor,—The report of the amazing results of your scientific researches has reached me, and I congratulate you most heartily on the originality and acumen which you have displayed in your investigations.

A new light has dawned upon our country. Instead of groping in the darkness of political warfare, ensnared by party ties and jealousies, the statesmen of the future will be able to calculate and determine the correct course with mathematical precision and perfect accuracy.

No one can dispute the truth of a proposition in Euclid, or the genuineness of Newton's laws; and if your method enables men to calculate and determine the correct political course of action, to solve political problems as easily as exponential equations, why—then adieu to the bickerings of party, the querulous complaints of the Opposition! Nay, joy to the Ministry!

There will be no Opposition! Our statesmen will be able to guide the great ship of the State by means of charts which know no error; and they will resemble an association of savants met together to determine the exact moment of the transit of Venus, or to examine the degree of density of a comet's tail.

This condition of Parliamentary procedure is much to be desired; you have shown how such an ideal state of things may be obtained. In the name of the Government I thank you for your endeavours on behalf of your country's welfare, and look forward to a further development of your admirably conceived system. As in the domain of ordinary science there are complex questions which defy the acumen of the philosopher;

so in polemical science there may be questions which present the same difficulties and complications. But as the first are daily yielding before the persevering attacks of the mathematician, so I doubt not polemical science will soon overcome the various problems which may arise.

But it is mainly on my own account that I venture to address you. I desire to consult you with regard to certain matters—political complications—which have recently occupied the attention of Her Majesty's Ministers. By the help of your new science, can you aid us in our deliberations? Of course, I am writing to you in *strict confidence*, and beg that you will keep this communication profoundly secret. I fear that would be a hard task for many of your sex, who do not possess your knowledge and powers of mind; but I have great confidence in your discretion.

These are the problems which are presented to us for solution:

1. Some members of the Cabinet are secretly in favour of Protection, and the country is rather stirred by the question. Can you, from your knowledge of the contact of curves and nations, help us to determine what course we ought to take with regard to Spain, for example? Are the principles of Adam Smith mathematically correct?

2. I observe that England is represented mathematically by an ellipse. Are we right in assuming that Ireland is a portion of that ellipse? Or, on the other hand, in our chart of nations, must we describe that troublesome country as a rotating parabola, or complex figure, altogether outside our more favoured State?

3. Do you consider, from your minute observation of our social system, that the form of our elliptical government is gradually undergoing a change, and that a revolutionary parabolic tendency is observable in the action of individual particles?

4. Is it not possible that the differences in the policy of the various nations of Europe; the difficulties which beset the carrying out of international law; the jealousies, quarrels, and rivalries of States might disappear, if the same form of government (*i.e.*, elliptical) were adopted in each?

If you will kindly favour Her Majesty's Ministers with your opinion on these questions, they will owe you a debt of gratitude, which they, as representatives of the nation, will do their utmost to repay.

With every good wish for your further success in the regions of polemical science,

I beg to remain,
My dear Lady Professor,
Your faithful servant[4]

[Editorial Note]—The next letter is not of quite the same pleasing nature as the foregoing, and shows that it is impossible to please everyone, even if that happy consummation were desirable. This letter was evidently called forth by some remarks which the learned Lady Professor had made in her third lecture with reference to eccentricity in dress.

Our readers will recollect that the professor pointed out that an extravagant 'bloomer' costume—half male, half female—was no more a sign of genius than æsthetic dresses, always betokened the artist.[5] This latter statement evidently gave great offence to the members of a society which called itself the 'Æsthetic and Dress Improvement Association,' and the following letter is the result of one of their solemn conclaves:

Oscar Villa, South Kensington,
June, 18—.

The Secretary of the Æsthetic and Dress Improvement Association presents his compliments to the Lady Professor of Girtham College, and begs to contradict emphatically her statements with regard to a subject upon which she is evidently in entire and lamentable ignorance, and to protest against her aspersions upon the artistic studies of this and kindred societies.

He begs to state that true æsthetes are *not* eccentric (they leave that to lady professors and her Philistine followers); that to dress becomingly is one of the principal objects of life, and that true greatness is achieved as much by the study of the art of dress as by any other noble pursuit or graceful accomplishment. Are not Horatio Postlethwaite, Leonara Saffronia Gillan, Vandyke Smithson entitled to greatness? And yet their laurels have been won solely by the art of dress.

Perhaps the lady professor has never read 'Sartor Resartus'! In conclusion, he would ask the Lady Professor to refrain from casting obloquy upon the work of the Association which he has the honour to represent; to prevail upon her pupils to abandon the unfeminine attire which some of them have assumed, contrary to the first principles of art; to array themselves in flowing robes of sage-green and other choice colours (patterns enclosed), and to study art, instead of absurd mathematics, which no one can understand, and do no one any good.

(Approved by the Committee of the Æsthetic and Dress Improvement Association.)
June, 18—.

[Editorial Note]—The next letter, written by a pupil of the Lady Professor, requires no explanation, and

speaks for itself.

Jesus College, Cambridge,

March, 18—.

My dear Tutor,

You will be glad to hear that after superhuman exertions I have at last succeeded in passing my Little-go, and I am eternally grateful to you for all you have done for me. I should never have got through if it had not been for you.

All the coaches in Cambridge would never have managed it, but you drove me through in a canter. And why? I never could make up my mind to work for them; but when I coached with you, you made me like it. I almost revelled in the Binomial when you wrote it out for me; and then I could not help listening to you; and you looked so grieved when I would not learn, and made me feel such a brute; so somehow or other you drove some mathematics into my head, and I pulled through.

By-the-bye, I think you must have tried the 'brain wave' dodge with the examiners, as five out of the six propositions in Euclid, which you told me to get up specially, were set! I wish I could read people's thoughts; can you read mine? If I were a Don, or a Fellow, or something, I would advise the University to have some lady professors like you to teach the men, instead of some of these sleepy old tutors. It would be a great improvement, and I am sure we should get through a great deal more work.

They have given me a place in the Jesus Eight, which I shall take now that I am released from your professorial ban, and have time for rowing. But I don't half like giving up mathematics. You see, I have grown fond of the study. Do you think you could make a wrangler of me? At any rate, I should like to come to your lectures again. May I?

Your Grateful Pupil.

Notes

4. It is to be regretted that this letter has evidently fallen into the hands of some autograph collector, who has ruthlessly cut off the signature; but the reader will easily determine, after careful perusal of the document, from whose pen it emanated.
5. Cf. page 36.

Paper V.

A Lecture Upon Social Forces, With Some Account Of Polemical Kinematics

Most noble Professors and Students of Girtham College,—Since last 'I wandered 'twixt the pole and heavenly hinges, 'mongst encentricals, centres, concentricks, circles, and epicycles,' like the great Albumazar, and found them full of life and wisdom for the guidance of our States and laws, I have turned my attention to the Applied Mathematics, in order to determine what other truths this shaft may yield.

The strength of all sciences, according to Bacon, consists in their harmony; and it is truly marvellous how perfect this harmony is, if our ears are tuned aright to hear it. We have observed how the beautiful and regular laws of curves and cones correspond to the social laws of States and nations, guiding them as if by word of counsel, admonishing them on what principle they ought to regulate their governments and inter-relations. We have seen that the laws which govern thought and light and sound are almost identical, and that harmony pervades not merely the ordinary sciences, but extends her benign influence over these newly discovered fields of scientific research, which I claim to have discovered.

All this may appear at first sight surprising; but the real philosopher, who knows that all kinds of truth are intimately connected, will receive such revelations of science with satisfaction rather than astonishment; for this new science, which has opened itself out before me, is only an extension of other well-known laws and discoveries which have come down to us from the remote past.

If my investigations should appear to you, most noble professors, somewhat novel and imaginary, remember the maxim of the sage, that in the infancy of science there is no speculation which does not merit careful examination; and the most remote and fanciful explanations of facts have often been found the true ones. Perhaps some 'self-opinionated particle' (I speak mathematically) may have been inclined to laugh at our theories and discoveries, as the wise fools of the day laughed at Kepler and his laws; but time has changed the world's laughter into praise, and a century hence our discoveries may rank among the achievements of modern science. As Cicero says, 'Time obliterates the fictions of opinions, but confirms the decisions of nature.'

I have not shunned, most noble professors, to enlist Imagination under the banner of Geometry; for I am fully persuaded that it is a powerful organ of knowledge, and is as much needed by the mathematician as by the poet or novelist. It is, I fear, often banished with too much haste from the fields of intellectual research by those who take upon themselves to give laws to philosophy. We need imagination to form an hypothesis; and without hypotheses science would soon become a lifeless and barren study, a horse-in-the-mill affair ever strolling round and round, unconscious of the grinding corn. In my previous investigations my imagination pictured the symmetry of curves and States; the hypothesis followed that the laws which regulated them were identical, and you have observed how the supposition was confirmed by our subsequent calculations.

In this lecture I propose to examine some of the forces which exist in our social system, and shall endeavour to estimate them by methods of mathematical procedure and analogical reasoning. We will begin with the old definition of Force as *that which puts matter into motion, or which stops, or changes, a motion once commenced*. When a mass is in motion, it has a capacity for doing work, which is called *Energy*; and when this energy is caused by the motion of a body it is called Kinetic Energy (in mathematical language $KE = \frac{1}{2} MV^2$).

Another form of kinetic energy is called Potential Energy, which is in reality the capacity of a body for doing work *owing to its position*. For example we may take an ordinary eight-day clock. When the weights are wound up, they have a certain amount of potential energy stored up, which will counteract the friction of the wheels and the resistance of the air on the pendulum. Or, again, we have the example of a water-wheel: first the water in the reservoir, being higher than the wheel, has an amount of potential energy. This is converted into kinetic energy in striking against the paddles, and after this we have potential energy again produced by the action of the fly-wheel.

By the principle of conservation of energy, if we consider the whole universe, not our planet alone (for its heat and energy are continually diminished to some slight degree), we find that *no energy is lost*.

Force is recognised as acting in two ways: in *Statics*,

so as to compel rest, or to prevent change of motion; and in *Kinetics*, so as to produce or to change motion; and the whole science which investigates the action of force is called *Dynamics*.

All this is of course pure mathematics, and I have made these elementary observations for the benefit of my younger hearers, the students of this University. My grave and reverend seniors will pardon, I am sure, the repetition of facts well known to them for the sake of those who are less informed than themselves.

Now before I proceed further, I will endeavour to point out that these elementary truths of physical science hold good in our social system. Each individual is a mass, acted on by numerous forces, capable of 'doing work,' which work can be measured and his velocity calculated. Some individuals have a vast *potential energy*; that is to say, from their position and station in the social system, they have a power which is capable of producing work which a less exalted individual has not. Like the weights in an eight-day clock, or the water in a reservoir, they have a capacity for doing work, owing to the position to which they have been raised. How vast the influence of a Primate or a Premier, a General or a King! And yet their power is chiefly potential energy, arising from the position they occupy, not from the individuals themselves. Schiller has described this in poetical language, which, strange to say, is mathematically correct:

> 'Yes, there's a patent of nobility
> Above the meanness of our common state;
> With what they *do* the vulgar natures buy
> Their titles; and with what they *are*, the *great*.'

Other forces may have raised these men to their exalted

positions; but their influence is due to their height, their potential energy. Placed on a lower level, they would cease to have that power. How calm the dignity of this potential rank! The water in the reservoir is scarcely ruffled or disturbed, as if unconscious of its power; when it has lost its force it rushes along with a sullen murmur and a roar, howling and hissing and boiling in endless torture, until—

> 'It gains a safer bed, and steals at last
> Along the mazes of the quiet vale.'

So the vulgar crowd rushes on, with plenty of kinetic force, making noise enough and looking very busy; while those who seem to sleep in calm forgetfulness, exercise their potential energy, and do the real work of turning the great engine of the State.

There are attractive and repulsive forces (more commonly the latter, the cynic will say) in our social system, but each individual is the centre of various forces acting upon him. In nature all matter possesses the force of gravity, and whatever the size of two particles may be, they mutually attract each other. The earth attracts the moon; the moon attracts the earth. A stone thrown up into the air exercises an infinitesimal force upon the earth; so in the social system every individual, however small and insignificant he may be, exercises some attractive force upon his neighbour. There is no one in the world who does not exercise some influence for good or for evil upon his fellows.

The force of *cohesion* is manifest in society as in nature, that force, I mean, which resists the separation of a body's particles. Different bodies possess different powers of cohesion, *e.g.*, the cohesion of chalk is far less than that of flint embedded in it; even the same body possesses different powers of cohesion in different directions, *e.g.*, it is easier to split wood in the direction of the fibres than perpendicular to them. If by our old principle of continuity we change the words 'bodies' into 'States' or 'individuals,' we shall see that the same laws hold good in social science as in natural philosophy.

These are a few analogous laws which I have taken almost at random; but it must strike the most casual listener to my remarks that it is wondrous strange that men, regarded as social beings, should possess the same qualities, and be governed by the same laws, as the rest of *matter*. As Bishop Butler says, 'the force of analogy consists in the frequency of the supposed analogous facts, and the real resemblance of the things compared.' It appeals to the reasoning faculty, and may form a solid argument. Hence, if we can prove the similarity of various laws and conditions, we may not be wrong in assuming by analogy the identity of those laws and conditions.

I have stated my case in this manner in order to convince the gainsayers, if any such there be, and to banish any doubts or questionings which may have arisen in your minds. I will now proceed with some further investigations, full of the most profound interest and importance.

Doubtless many of the lady-students present are in the habit of welcoming peaceful evening in with a potent draught of 'the cup which cheers but not inebriates;' and as men are great flatterers (for imitation is the greatest flattery), I believe the male portion of my audience have been known to follow that excellent example. Some perhaps are in the habit of burning the midnight oil, and keep their eyes open by means of this fruit of the hermit's pious zeal, endowed by high omnipotence with the power of hindering sleep;[6] but that practice I do not advise, as that delicate portion of our system, the nerves, especially of women, often becomes injured by such stimulating doses.

However, you will have observed (if you do not follow the modern pernicious fashion of taking tea without sugar) that numerous bubbles are formed upon the surface of the liquid. After a few moments these unite into one central mass of bubbles by the force of mutual attraction.

It appears from considerations which are detailed in works on physical astronomy, that two particles of matter placed at any sensible distance apart attract each other with a force directly proportional to the product of their masses, and inversely proportional to the square of their distance.

Now, suppose that we have a number of circular masses situated upon a plane surface, they will attract each other with a force which may be determined with exactitude; and the greater the masses the greater the force. We will now apply this to polemical science.

The agricultural settlement is the first stage in the civilization and formation of a State. How did this arise? First, a single family immigrated to some uncultivated parts of the country, perhaps accompanied by others, who formed a little colony. Other settlements were made in other parts of the land; and thus the country became overspread with these detached and separate communities.

An eminent writer declares that these settlements can be traced in the beginnings of every race which has made progress; that they were characteristic of those races in Greece and Italy, in Asia and Africa, which grew into the opulent and famous cities in which so much in the early history of civilization was developed. The colonies of England have been formed in the same way, just as in olden time England itself was occupied when the Roman power ceased.

These settlements correspond to the circular masses

situated on the plane surface; they were quite separate from each other, each having its own laws, its own headman or ruler, its own assembly or parish council. But as time elapsed, the force of mutual attraction set in; by degrees these separate settlements were drawn together by force which increased in proportion as the settlements increased; until at last one united kingdom was formed under one king, governed by uniform laws and regulations. The bubbles have blended, the circles have come together, and one large circle or other curve is the result. This may be called the *Law of Social Attraction*. In accordance with the results of one of my previous lectures, I have taken the circle as representing the simplest form of government, which figure, in the case of the elementary settlements, must have been small.

Many of you, most noble professors, are doubtless accustomed to make experiments with the microscope. I will suggest a simple one, which illustrates very forcibly what I am endeavouring to show you. Take some particles of copper, and scatter them at intervals over the surface of an object-glass, and pour some sulphuric acid upon the glass. Now, what is the result? A beautiful network of apparently golden texture spreads itself gradually over the whole area of the glass. Steadily it pursues its way, and the result is beautiful to behold. The minute particles of copper were the original settlements scattered over the land; the sulphuric acid the civilizing agent; and the final picture of a united civilized homogeneous nation is well represented by the progressive and finally glorious network of gold. This example is of course outside our present subject, but it serves as a beautiful illustration.

As an instance of the attractive force exercised by small communities upon each other, I may mention the united kingdom of Germany, which is composed of numerous small

States and nations, which have been drawn together by the power of mutual attraction. Until recently they were each self-contained, separate constitutions, with their own kings and forms of government; but the attracting force, assisted by forces from without, has proved too much for them, and the great and powerful united kingdom of Germany is the result.

But why, you may ask, have not the people in Hindustan united in the same way? There the agricultural settlements remain as they did ages ago; separate petty chieftains rule under the all-governing power of England. Why have they not united?

To this objection I reply that there is in social science, as in Nature, a *vis inertia;* that is to say, there is a tendency in matter to remain at rest if unmoved by any external agency, and also of persisting to move, after it has once been set in motion. The *vis inertia* of some bodies is greater than that of others, and depends upon their weight and density. Now it so happens that the moral *vis inertia* of the Hindustani is very great, hence their tendency to amalgamation is small. They remain in the state in which they happen to be.

On the other hand the inertia of Englishmen is small, of Englishwomen smaller, and therefore their power of combining is greater. Here let me observe that the quality of inertia is one which ought to be removed as far as possible from each social system. Inertia was regarded as a capital crime by the Egyptians. Solon ordained that inert persons should be put to death, and not contaminate the community. As savages bury living men, so does inertia practise the same barbarous custom upon States and individuals.

Observe the putrid state of inert water, the clear and sparkling beauty of the moving stream, bearing away by the force of its own motion aught that might contaminate it.

Men more often resemble the stagnant water than the rivulet. A healthy social state enforces labour by natural laws, and banishes inertia as much as possible from the system. If the principles of some noisy English politicians were fully carried out, and all things made *'free,'* inertia would be increased, and listless indolence pervade the masses of our countrymen. I may say that inertia is not entirely unknown in our sister University of Cambridge.

The existence of social forces is supported by the testimony of Dr. Tyndall, who plainly recognises their power, though he does not attempt to expound their origin. 'Thoughtful minds are driven to seek, in the interaction of social forces, the genesis and development of man's moral nature. If they succeed in their search—and I think they are sure to succeed—social duty would be raised to a higher level of significance, and the deepening sense of social duty would, it is to be hoped, lessen, if not obliterate, the strife and heart-burnings which now beset and disguise our social life.'

I accept with gratification Dr. Tyndall's conclusions: to determine, examine, trace, calculate these social forces which exercise such a powerful influence on our characters, our lives, our customs, which produce the greatness of the State, or drag it down with irresistible strength from its pinnacle of glory to an abyss of degradation; to estimate such forces is the great and noble object of our lectures and researches in this University. Prosecute, most noble professors, your studies in this direction with all the energy of your enlightened intellects, and there is yet hope that this new science, which I have endeavoured to sketch out, however feebly, may be the means of saving our beloved nation from degradation and ruin, and raising her to a higher level of glory and honour. I hope to continue the subject of social forces in my next lecture.

Notes

6. A Chinese legend relates that a pious hermit, who in his watchings and prayers had often been overtaken by sleep, so that his eyelids closed, in holy wrath against the weakness of the flesh, cut them off, and threw them on the ground. But a god caused a tea-shrub to spring out of them, the leaves of which exhibit the form of an eyelid bordered with lashes, and possess the gift of hindering sleep.—Dr. Ure.

Paper VI.

On Social Forces *[continued]*— Polemical Statics And Dynamics

Most Noble Professors and Students of Girtham,—We have embarked upon a stormy sea of speculation, on a voyage of grand discovery, and the dangerous waves of adverse criticism, and the deceptive under-current of prejudice, often make the steersman's lot by no means an enviable one. But our vessel is sound and perfectly equipped, and therefore I do not fear to guide her across the great unknown.

It may have occurred to you that the problems which present themselves for solution in social science are far more difficult and complicated than those which arise in ordinary mathematics. That is undoubtedly the case; but this extra degree of difficulty is due to the fact that we make no assumptions; we take the things as they really are, not as they are assumed to be.

In physical science, if we take into consideration the resistance of the air, the curvature of the earth, the rigid connection which exists between particles in the same body, and a host of other things which are often conveniently neglected in elementary works, how complicated the various problems become! So we must not be surprised at some of the difficulties which occur in social science, as nothing is neglected; the whole problem is before us, and having solved it we need not make allowances for any falsely assumed *data*.

It is possible that other professors of this science may come to slightly different conclusions to those which I have arrived at. That is only to be expected, because their original observations may have slightly varied. But in physical science allowances are made for different observers. In astronomy, for example, we find the value of the 'Personal Equation.' One observer on looking through the telescope may take the meridian of a star rather differently from another watcher of the heavenly bodies, and the *personal equation* is used to make allowances for this quickness, or slowness, of observation.

So in social science there must be a personal equation too, and our object ought to be, in the ordinary affairs of life as well as in the higher duties of scientific action, to make our personal equation as small as possible. But until the old proverb, '*Quot homines, tot sententiæ,*' has ceased to have any meaning, there will be abundant need of this most useful aid to accuracy.

The close connection which exists between social forces and material forces is plainly shown by the doctrine of the conservation of energy. 'This doctrine,' says Dr. Tyndall, 'recognises in the material universe a constant sum of power made up of items among which the most Protean fluctuations are incessantly going on. It is as if the body of nature were alive, the thrill and interchange of its energies resembling those of an organism. The parts of the stupendous whole shift and change, augment and diminish, appear and disappear; while the total of which they are the parts remains quantitatively immutable, *plus* accompanies *minus*, gain accompanies loss, no item varying in the slightest degree without an absolutely equal change of some other item in the opposite direction.' So do the forces in the social world ebb and flow, rise and fall, carrying on the same universal law which regulates the energy of material force.

I will now proceed to enumerate some of those forces which exercise such a powerful influence on society.

First, let us take the force of *Public Opinion*, which seems to exercise a relentless sway over the minds and manners of men. This is a very subtle and secret force, which is most difficult to trace, and resembles electricity in the science of physics. We cannot see it, but are only able to judge of its power by its results. Its point of application is not in the individual, but in the collection of individuals who make up the social system; and it is, in reality, the resultant of, or the compromise between, the various elementary forces which make up human society.

Yes, compromise is a purely mechanical affair, based on the principle of the parallelogram of forces; and as public opinion is the result of a compromise, we may calculate its force. For example: 'It is required to know the state of public opinion in the matter of politics, when the results of a General Election show that the Conservatives are to the Liberals as 10 : 9.'

Let OC be the direction of the Conservative force.

Let OL be that of the Liberal.

Then by *data* OC : OL :: 10 : 9.

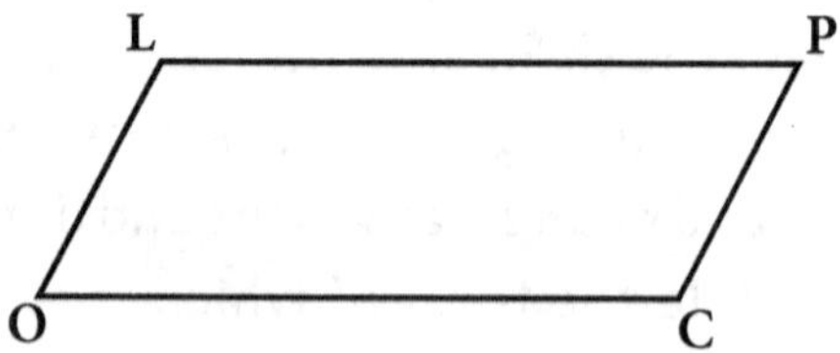

Complete the parallelogram, and join OP.

Then OP represents the force of public opinion in magnitude and direction.

N.B.—The direction of OL is determined by the amount of deviation of the policy of the Liberals from that of the Conservatives.

As in physical, so in social science, impulsive forces sometimes act, and effectually disturb our system and our calculations. Public opinion is very liable to the action of disturbing forces. Panic is an impulsive force, which defies the power of the most learned professors of social science to determine its magnitude and direction.

Some strange unforeseen catastrophe—the fascination caused by a brilliant and unscrupulous orator, a cruel wrong, a blind revenge for real or imaginary injustice—will sometimes rouse one element of passion latent in the vast body of public opinion; so that it breaks with all that hitherto restrained and balanced it, and precipitates society into a course of conduct inconsistent with its former behaviour, and bloodshed, revolution, the breaking-up of laws, are the terrible results of panic or revengeful passion.

Society is, as it were, split up by the terrible action of such impulsive forces, just as wood is split up by the repeated blows of the hatchet. It is, therefore, the duty of statesmen to increase the power or force of cohesion, to strengthen the fibres of the State, so that the force of such impulsive blows may not be felt, nor disturb the continuity of the framework of the State. If such measures had been adopted in the neighbouring country of France, much misery might have been avoided, and the terrible revolutions which have so frequently convulsed her social system entirely prevented.

Friction is another disturbing element in our calculations, and although it may be made a useful servant, it is a bad master in mathematics, as in polemics. Without the aid of friction, progress would be impossible. For example: Take the case of a man with perfectly smooth skates on perfectly hard, smooth ice; he would be unable to reach the land unless he had provided himself with some stones, by throwing which he would just be able to get to his destination by a backward motion. The engine would be unable to proceed on its iron road if it were not for friction.

The same is true in polemical science: the government of the country would not be able to be carried on under our present conditions if it were not for party friction. But suppose it increased indefinitely, *party friction* becomes party *obstruction*; and the engine of the State would no longer proceed smoothly and evenly along its appointed course at the rate of sixty miles an hour, but would resemble an old-fashioned coach, up to its axle-trees in mud, its motion altogether stopped by the action of party friction.

We have seen that forces have two ways of acting: that of compelling rest and that of producing motion. In statics forces act so as to prevent any change of motion, or disturb the body's original position. In kinetics, on the contrary, the power is recognised as acting so as to produce or change a body's motion.

Now, in polemical science we have these two ways of considering the action of forces. There is the *statical* or *conservative* force, which compels rest, which seeks security, stability, and peace, and is not ardently devoted to change. It reduces the system to equilibrium.

There are, of course, two kinds of equilibrium—*stable* and *unstable*—according as the social and political system is in a healthy or unhealthy state. If a body is in stable equilibrium, and any slight motion takes place, the body will return immediately to its former position; but if in unstable, it will decline further and further away from its original position, and be entirely upset.

So a healthy and sound conservative equilibrium is not disturbed by outside forces, and the State will resume its former position of stability and rest when the opposing force is withdrawn. But an unhealthy and insecure conservatism is as easily disturbed as an egg balanced on its narrow end.

The kinetics of society, that is to say the Radical way of estimating force, is the party of motion, generally supposed to be the 'party of progress.' It has therefore many attractions in the eyes of those who delight in motion, speed, and rushing about. To run at full speed, to feel the keen air upon one's face, to experience the delightful sensation of freedom of will, and limb, are joys which cannot be denied.

Such exercise is beneficial to the system, bodily or political. Motion is the life of all things; it is characteristic of nature; it adores nature; because it is an emblem and characteristic of life. The ceaseless rolling of the ocean waves, the swaying of the trees, the bending of the flowers, the waving of the corn, all these fill us with pleasure; whereas a flat uninteresting plain, unrelieved by the motion of terrestrial objects, is depressing to the spirit. So there is much to be said in favour of motion, and Carlyle has defined progress as 'living movement.' And men love this 'living movement,' and take up the Laureate's cry:

> 'Forward, forward, let us range,
> Let the great world spin for ever down the ringing
> Grooves of change.'

But, after all, there is a danger in this everlasting motion. We cannot tell whither this progress may lead. It may be along a safe sure road; but perchance a precipice may open out before us; and rejoicing in the acceleration of our velocity, with eyes intent upon some distant heights of glory and ambition, we may not discover our danger until it is too late to stop, and a terrible plunge into an unknown abyss of turmoil and tumultuous waves is the alarming result of an unguarded policy of unrestrained 'progress.' I recall to my mind the quaint words of Holmes which aptly illustrate my contention.

> 'If the wild filly, "Progress", thou would'st ride,
> Have young companions ever at thy side;
> But wouldst thou stride the staunch old mare,
> "Success,"
> Go with thine elders, though they please thee
> less.'

Progress and success do not always go together hand in hand; and while motion is essential to life, it is not always safe to urge a country forward at too great a speed; and security and stability are quite as important to the nation's life as actual progress.

There are other impulsive forces which act occasionally in the sphere of politics, and which baffle all our calculations, and exclude scientific considerations of the polemical problems which arise. *Ambition* is such an impulsive force, and when the rulers of the people are actuated by it, and struggle for money, place, and power, politics is degraded from its position as a science, and it becomes impossible to estimate the result of forces so generated.

In my next lecture I propose to treat the important

subject of the Laws which govern States and Governments, and which regulate, generate, and control the social forces which we have seen at work in the body politic.

Paper VII.

Laws Of Political Motion

Since the last time I had the honour of addressing you on polemical matters, I have met with a passage in the writings of M. Auguste Comte which afforded me much pleasure. It seemed to be the one word for which I had been waiting, and confirmed many of my own impressions and speculations. He lays down two propositions: first, that the constructive politics of the future must be based on the history of the past; and second, that political science is a composite study, and presupposes the complete apprehension of every branch of science, beginning with the physical, such as astronomy, and ending with the moral, such as ethics and sociology.

M. Comte evidently does not regard as a vain dream and imaginative speculation the theory that it will be possible for statesmen to calculate a policy, and to determine a course of action by purely scientific considerations. May I entertain the hope that in this university, where all branches of physical science have found a home, and are studied by most able and learned professors, the science of politics may be pursued under most favourable circumstances? I trust that each professor will bring before me the results of their deliberations, and contribute to the growth of this particular science for which our university has already become deservedly famous.

My present lecture is devoted to the important consideration of *Law*. At first sight it may appear to you that the wills and passions of mankind are so diverse and unknowable, that it would be absurd to suppose that they can be calculated, or rendered amenable to any law. But Professor Amos has pointed out that in proportion as we examine history, and compare the actions present and past of different nations and states, the more uniform does human nature appear; the more calculable the actions, sentiments, and emotions of large masses of people. As we have already stated, the difficulties of the study are not likely to deter the professors of Girtham College from the pursuit of any particular branch of science.

A priori we might suppose from analogy that these polemical laws existed, as there is no department of nature which is not governed by law. It is an essential feature in nature, and also in government. What is political economy but the study of certain laws of nature? These were first discovered by Adam Smith, and have since been traced and estimated by such men as Ricardo, the two Mills, Professor Cairnes, Jevons, and many others.

Moreover, our physical constitutions are governed by laws, which physicians have determined, and which it is perilous to resist. Our moral constitution is also governed by laws, which evidently exist, although it is difficult to find them out. But the nation is only an assemblage of individuals; and since individuals are so governed, it is only natural to suppose that the nation, composed of individuals, is so constituted and controlled. And not only is that true, but we shall see that polemical laws are as permanent and universal, as invariable and irreversible, as the laws of nature which regulate the courses of the heavenly bodies, and raise the tides, or depress the sandstone hills.

We may notice first the preponderant impulse observable in a nation's life in favour of supporting existing facts and institutions; and every reformer has discovered the difficulty and danger of changing or opposing the customs and habits of the people. As a wheel will travel most smoothly along a well-worn groove, whereby friction is diminished, so there is a natural national tendency always to run along those paths with which the habits and customs of the people have made them familiar.

This law is nothing else than Newton's first law of motion, which is quite as applicable to human masses as to lifeless matter. The tendency of matter to remain at rest, if unmoved by any external agency, and of persisting to move after it has once been set in motion, is a conservative tendency; and is as true in political science as in any other.

The special branch of our science, which we may call the *Biology of Politics*, shows how absolute is the domain of law in polemical matters. The law of human life is that men are born, grow, become strong and vigorous, and then decay and die. This is the law of life, to which we must all yield an enforced obedience.

This same law is observed to be at work in the heavenly bodies; and astronomy shows us that planets are born, flourish, and at length die, just as our human bodies do. The moon is, as you may have observed, a dead planet, such as our earth may be some day.

The same growth and decay are also manifest in national life. First, there is the birth of the nation, which sometimes lies a long time in a dormant state, and then wakes up to life and energy. China and Russia are examples of dormant States, just waking from a long sleep of childishness and ignorance. The next stage is the strong an healthy period of its existence, which England is at present enjoying; and then, after various stages of gradual decline, we come to the senile period of national life, when every energy and faculty, every national feeling and power of invention, are completely exhausted. As an example of this depressing condition, we may mention Turkey and several of the effete States of South America.

Sometimes, when life is nearly extinct in the human body, physicians have made use of the power of galvanism, in order to revive the dying energies. This process of galvanizing a State into life was tried by Lord Palmerston and others on the worn-out frame of Turkey. But such attempts can only meet with partial and transitory success; and where the loss of national power and faculty betokens the senile period of the nation's existence, it is vain to attempt to restore its former life and energy. The study of the biology of politics presents many interesting and important details in this special branch of knowledge; and I commend this part of our subject to the special attention of the professor of physiology.

The law of development is observable in nations as in nature. Recent scientific discoveries have tended to take away all ideas of *chance* in the workings of nature, and have substituted *law* instead of it. It would be unscientific and incorrect to speak of the world being formed by the 'fortuitous concourse of atoms.' So we cannot speak of a State being generated in this manner. Laws—economical, geographical, natural—preside over the formation of States and nations, and produce their further development.

The laws of political motion occupy the same prominent place in our new science as Newton's laws do in ordinary dynamics. These are very important in calculating the positions which various States will occupy in the future.

First, we have the *doctrine of nationality*, which prevented the progress of Austria into Italy, and of the Bourbons in Naples, and produced the amalgamation of the small German States in the great empire of Germany.

The second law of political motion is the doctrine of the *independence* of all true States, and the equality of all States to each other. This had its growth in feudalism; and all the chief wars of modern times have been the result of the efforts of nature to establish this law of independence. The doctrine of intervention is a modification of the preceding law, and is applicable when the law of necessity demands its use, such as the restoration of order after protracted anarchy, the abolition of slave trade, etc.

The third law is the *law of morality*. Just as for each man there exists a *right* and a *wrong*; just as *duty* and *conscience* are certain elements in his daily motion, which dictate his course of action, although he may choose to neglect them; so a nation is bound by the same moral laws which govern the individual; and a nation errs if it transgresses them.

Christianity is the agent which has produced so powerful an influence in making men obey the dictates of conscience and walk in the path of duty; and I read with thankfulness the conclusion of Mr. Amos, that Christianity has triumphed quite as much in moralizing secular politics as it has in the sphere of individual life.

These are some of the principal laws of motion which I have observed at work in various States and nations. Inasmuch as political science embraces, in addition to the physical sciences, all those branches which are contained in ethics, economics, jurisprudence, sociology and others, the laws of each are generally applicable to the whole grand subject of which my lectures treat.

Other general laws may be deduced, and have been enumerated in my previous lectures, from the social properties of curves and conics; and when our researches are complete we may hope to produce a code of laws for the guidance of our statesmen which maybe of immense use in determining the policies of the future. Already there is strong evidence that the affairs of this country are being conducted on sound scientific principles, rather than by any species of guess-work or haphazard contrivances.

The use of history is recognised as extremely important in determining a future line of conduct; and statesmen are in the habit of endeavouring to find from their study of the past what is the logical sequence of events. Just as mathematicians endeavour to determine the law of a series of figures, and having found the law, can write down the next, and the next, *ad infinitum*; so scientific politicians may be able soon to establish the various laws of a series of events, and calculate their course of actions. That there is considerable progress in this direction is manifest by the value which they place upon statistics, and their continued use of this important information.

There are a few great evils in our present system which are strongly opposed to any scientific methods in politics; and in the interests of the country as well as those of science they ought to be removed.

One great evil is the want of political and scientific knowledge on the part of the electors, who are in the habit of choosing their representatives on personal grounds, or party considerations, rather than on sound principles of political science. All this is opposed to any idea of law. Owing to the ignorance of the electors they fall an easy prey to adventurers and unprincipled politicians, who make all kinds of specious promises, tempt them with all manner of baits, and make self-interest instead of the welfare of the

State the principle of voting.

Selfishness is the ruin of social life and intercourse, the destroyer of all happiness, peace, and mutual trust in family life or in society. It is the root of most of the faults, vices, and crimes in the individual; and who can tell the endless disasters which will befall the State, where selfishness is the chief motive-power of the electors and the elected? A selfish statesman, one who goes into Parliament to gain his own ends and forward his own personal interests, is a disgrace to society—

> 'Feeling himself, his own low self, the whole,
> When he by sacred sympathy might make
> The whole one self. Self, that no alien knows!
> Self, far diffused as fancy's wing can travel!
> Self, spreading still, oblivious of its own,
> Yet all of all possessing!'

I have said that the ignorance of the electorate makes them an easy prey to such men; and until they have learnt to detect the false from the true, until they become acquainted with the elements of political science, and have been taught that their own selfish interests are not the highest aims of social government, it is vain to hope for a reasonable method of regulating the affairs of the nation, based upon logical laws and scientific principles.

And how is this work of educating the electors to be accomplished? Not, I maintain, by furious speeches and rhetorical displays; not by bribery, baits and banter; but by patient, never-ceasing labour, by lectures on history and science, by individual instruction, is the great work to be accomplished upon which the security and stability of the country depend.

Then we may hope that the 'Reign of Law' in polemical science may be ushered in with the joyful acclamations of an enlightened and united people, and its benign influence extend from the throne of the monarch and the council-chamber of his ministers to the hearth of the cottager.

Politicians will rule by law; policies be calculated by laws; people vote by law; and then methinks I see in my mind (to use the words of the blind old poet) a noble and puissant nation rousing herself like a strong man after sleep, and shaking her invincible locks; methinks I see her as an eagle, renewing her mighty youth, and kindling her undazzled eyes at the full mid-day beam; purging and unsealing her long-abused sight at the fountain itself of heavenly radiance; while the whole noise of timorous and flocking birds flutter about amazed at what she means.

Such is the glorious vision of the 'Reign of Law.' Let it be the business of every Englishman and Englishwoman to arrange the framework of our social and political system, that law may have an uninterrupted sway; then shall we be a united, prosperous, and contented people, and the reign of lawless agitators, bribery-mongers, and counterfeit states-men will have passed away into the oblivion and obscurity of a more suitable but less favoured region.

Paper VIII.

On The Principle Of Polemical Cohesion

In my previous lectures I have had occasion to mention the principle of cohesion; but it plays so vital a part in the constitution of States and their relations to each other that I consider it advisable to devote this lecture entirely to it.

This is a large and comprehensive subject, and embraces such principles as the Centralization of States; the Co-operation of States; Monogamic Marriage; Unions; Free Trade, and many others equally important. We have already noticed that cohesion is a well-known property of matter; that its influence is not confined to the regions of physical sciences; and that it is the manifest duty of all governments to increase the force of cohesion.

Various methods have been tried to accomplish this purpose. The principle of Feudalism was one of the earliest attempts to produce the cohesion of the nation; and, in an elementary condition of society, it was partly successful. The theories of 'Divine Right' and 'Social Contract' were other methods which have been adopted; and the unity of the Christian Church has been the great means of producing the cohesion of the State in olden times; and its aid may be again required for the same beneficent object in future complications and social disruptions.

But it is always advantageous in scientific pursuits to go back to first principles; and we will adopt that method in our present investigations. The social unit is the family; the multiplication of families makes the tribe; the multiplication of tribes makes the State; and, therefore, we shall not be far wrong if we consider the family tie as the first principle of political cohesion. I am in agreement with several learned thinkers upon this subject when I say that marriage is a most important political factor; and as marriage cannot take place without women, it is evident that women play a very important part in promoting the cohesion of the State.

This prominent position was duly assigned to women by one of our greatest political philosophers, M. Auguste Comte, who strongly opposed the fatal fallacy of ancient political systems, which greatly overestimated the powers of men, and depreciated those of women.

If the superiority of bodily strength be the sole cause of greatness in political and intellectual pursuits, then, most noble lords of creation, we yield to you the palm—you are our masters in this respect. But if, on the other hand, it can be shown that physical strength is not a requisite for great achievements in these occupations; if the powers of endurance, elasticity, adaptability, nervous energy, and patience are quite as needful as mere animal strength; then we women are quite as capable, and indeed more capable than men, for achieving political greatness.

In the 'good old days,' when the law of might was right, and the strongest arm was the most powerful machinery in the government of the country, women were compelled naturally to occupy a less prominent position in the conduct of the affairs of the nation; and for centuries they have been degraded by a dominating tradition, and supposed incapable of performing duties for which they were mentally well suited.

But those militant days are past. Animal strength and brute force are no longer needed in the councils of the nation; and the time has arrived when women should cease to be oppressed by the disparaging, illogical deductions of former generations, and when their assistance ought to be invoked in the great work of promoting the nation's welfare.

I have stated that marriage is an important political factor; and, therefore, women have always occupied a primary, though obscure, part in political affairs. The cohesion of the State has been produced by the secret influence of family life.

But it may be asked, What kind of marriage is most conducive to national cohesion? This question has been carefully and conclusively answered by a learned scientific writer, who shows that polygamic marriage never exists in an advanced state, as instanced by the history of Judaism and Mohammedanism; that a strict form of monogamic marriage is essential to political greatness and true progress in civilization. The cohesion of the State is destroyed by polygamy, and by any system which relaxes the binding nature of the marriage tie. 'Domestic disorganization is a sure augury of political disruption.'

Cohesion, the essential property of all rightly constituted nations, is often in danger of being lost when the State is geographically very large, or when local interests have greater power than the attractive force of the central government. To obviate this evil, the method of centralization has been adopted with satisfactory results, as in the case of the United States of America, and Germany.

By this means the local authorities are brought into close relationship with the central head, and the centrifugal influences of independent interests and customs are counteracted by the force of central attraction. Centralization increases the importance of the whole body, and, like the pendulum of a clock, regulates the movements of the whole State.

In some cases it tends to make the government despotic, when the local governments are entirely under the control of the central; and every enactment, and scheme, and plan checked and supervised by the chief officers of the State. Such was the system adopted in France by Napoleon III. But cohesion without the enforcement of a hard and rigid connection, a general supervision without severe tyrannical jurisdiction, are the best methods of securing the unity of composite States.

But the force of cohesion is evidently at work in the nation apart from centralization. Men who have a community of interests unite together for the purposes of strength and mutual assistance. They combine for the sake of securing means of support in sickness, and form benefit societies, such as the Order of Oddfellows or Foresters. This force of cohesion has produced trade unions, and similar institutions which exist for the purpose of protecting a common interest, and giving expression to the concurrent opinions of the members. These have their legitimate use in every civilized State, in spite of some of the disadvantages which follow in their train.

There are, of course, opposed interests in every community: *attractive* forces, which produce trade unions, guilds, corporations, companies, and the like; and *repulsive* forces, which result from the opposed interests of employers and employed, landlords and tenants, and similar pairs of different classes in the community.

As time goes on, and the State advances with it, these

forces will gain in strength; the cohesion of classes will become greater; association will grow as naturally as the bubbles form on the surface of our evening beverage. It is a law of nature, and therefore cannot be resisted. But the repulsive forces will be no less strong, and to calculate the resultant of these contending interests will be the problem for practical statesmen to solve.

The force of cohesion is also evidently at work, not only in individual States, but also amongst the nations of Europe, and of the world. That is to say, there is an evident desire for co-operation on the part of those nations who have attained to the highest degree of civilization and internal cohesion. International law is based on the principle of cohesion, and every day it is gaining power and favour in the eyes of our leading statesmen.

The doctrine of Free Trade, which, if universally adopted, would be of the greatest service to mankind, results from a desire for co-operation; and whatever evils may result from one-sided Free Trade in this country at the present time, there can be no doubt that ultimately the complete system will be adopted.

Sad is the fate of a nation when the force of cohesion is weakened. The first revolution in France is a proof of this assertion; there was no cohesion, no common faith, or loyalty to the throne and Government; and indeed the Government, which was rotten to the core, was hardly likely to awake any feelings of loyalty and respect; and therefore the social disruption which followed was only a natural sequence of events, and was prophesied with the accuracy with which an astronomer can foretell an eclipse. But that is not all; when the cohesion of the State is destroyed, it takes a long time to restore the action of the force; and, as in the case of France, further disruption is sure to take place.

In this lecture I have already enumerated some of the ways in which this force acts; there are doubtless others which will suggest themselves to you. But I contend that the prosperity of the State, and the peace of the world, depend upon cohesion.

Let this be your work, most noble professors, to promote the action of this helpful and life-giving force. Promote, as far as in you lies, the sacred union of family life. Encourage the generous feelings of true loyalty and patriotism amongst the people of this realm of England; counsel our statesmen with regard to the primary necessity of national cohesion, and the advantages of international co-operation; and your work will be blessed; your names will rank with those heroes of the sword and of the pen who have raised our beloved country to her present pinnacle of greatness and prosperity; and your memory will live in the hearts of your grateful countrymen.

Extracts From The Diary Of The Lady Professor

[Editorial Note]—We regret to state that the various MSS in the sealed desk are nearly exhausted, and are therefore compelled to present the series of lectures on polemical studies in an incomplete form. But we had the good fortune to light upon a brief diary which discloses some interesting information with regard to the Author's life and occupations. We append a few extracts:

June 3rd.—Arnold called again to-day—the fifth time during the last fortnight! His attention is rather overpowering, and wastes much of my valuable time. He says he hates science—the heathen!—and wants me to lecture in classics. He affirms that mathematics are dry and hard—too hard for women, and tend to make them unsympathetic and critically severe. I am afraid I was rather severe with him. But really he is very trying, and always seems to talk like a Greek chorus in the most profound platitudes. Arnold is a classical tutor at Clare College. My old pupil is getting on famously. Poor fellow! He seems quite oppressed with his work. But he is making great progress, and sticks to his books like—a student of Girtham College!

June 4th.—Lectured on the Scientific Basis of Blackstone's Commentaries; afterwards received pupils until 1 p.m. Really Blanch S—— is more tiresome than ever. It appears that she has taken up with a young undergraduate of King's, and there is no prospect of any improvement in her work unless this nonsense is terminated. How foolish some of my sex are, in spite of their improved opportunities! I blush for them! Arnold has sent me a copy of Robert Browning's 'Belaustion,' in order to make me like classics, and give up science. Misguided young man! He has written some tolerable verses on the fly-leaf; but I have no intention of playing Belaustion to his 'entranced youth.' These are his verses:

> 'My lady dear, if I may call you so,
> For you are dearer than all else beside,
> I know the love you bear to golden verse,
> To golden thoughts enshrined in classic lore,
> To all that's beautiful; so here I send
> Some echoes of the songs of ancient days,
> Attuned and chanted by an English bard,
> Who fires one's old love for the rolling lines
> Of youthful Hellas; may your cultured ear
> Receive, and gladly welcome his sweet song.
> And while we revel in the poet's dream,
> And hear his actors speak, we'll play our parts.
> You, sweet Belaustion on the temple-steps,
> Taking your captors captive by your voice;
> And I, the youth who, more entranced than all,
> Was bound by fetters that he would not loose;
> And so we'll play our part. What say you, dear?'

     P. Hampson

June 6th.—Have just seen our new Professor of Physics, Amelia Cordial, who is an excellent woman, and well suited for the high office which she holds. She has told me of the foolish conduct of Lady Mary, who is evidently of opinion that the professorial mantle ought to have fallen on her shoulders. Really, this jealousy in the ranks of the learned is most disgraceful; and the bickerings which arise from disappointed ambition, the envyings and silly quarrels, are the weak places in our female collegiate system.

Such good news! The wrangler list is just out, and my hard-working pupil is *bracketed twelfth*! This is really delightful, and abundantly repays us for all our hard toil. But really I have not found working with him distasteful; he is such an excellent pupil, so painstaking and eager, that I have quite looked forward to his coming, and found him much more interesting than some of these foolish maidens. But I almost dread seeing him. He will be so elated and overpoweringly grateful, whereas I ought to be grateful to him for all his work for me; for I am sure he would never have gone in for the Tripos if I had not persuaded him. Well, I wonder why he does not come to tell me of his triumph?

June 7th.—It has come! and I half expected it. My eager pupil writes with all the energy and love of his noble nature to ask me to be his wife! He says *that* is all he cares for, and only values his Honours as a step to a higher honour and dignity, that of gaining my love and being my husband. All this is very nice to read; but a terribly difficult problem is placed before me for solution. I do indeed love this dear, good fellow—no one could help doing so, I am sure; but do I not love science more?

There is a stringent regulation in this University that no one shall occupy the position of professor who is bound by any domestic ties or cares. All married women are excluded.

If I say 'Yes,' I must resign my high position, leave this beloved college, give no more lectures to entranced audiences. In the interests of science, ought I to refuse, and sacrifice my heart's affections for the cause of mathematics? But if I say 'No,' I must give up—*him*; *sacrifice* his happiness too, and blight his life. Was ever anyone so perplexed? Science, aid thine obedient servant! May I not determine this vital question by thine all-pervading light?...

Conclusion

[Editorial Note]—We had just arrived at this exciting moment in the life of the learned and accomplished lady whose writings form the subject of these pages—a moment when love and science were trembling in the balance—when a footstep was heard upon the stairs leading to our study, and ere we could secrete our MS. the door was opened, and a well-known voice exclaimed:

'I do not know why you should have become so studious lately, Ernest, and why you should refuse to take me into your confidence. You spend hours and hours in this room all by yourself, writing away, and never say a word to me about the subject of your literary work. There was a time when things were different, and you were not so slow in availing yourself of my help, and asking my advice.'

We murmured something about taking up the pen which had been laid aside by a far abler hand, and our deep gratitude for past assistance in our work, which could never be forgotten.

'And do you think that I cannot help you now?' our visitor replied, in a very injured tone of voice. 'Is the old power dead, because it has not recently been used? Ernest, I think you very ungrateful not to confide in me. Come, tell me what you are writing.'

A suggestion about the proverbial curiosity of women rose to our lips, but died away without utterance. In the meantime, her eyes wandered over our study-table strewed with papers, and lighted upon the well-worn desk.

'Why, Ernest, where did you find this? My dear old desk, which has been lost ever so long! I do believe you have been ransacking its contents! Why did you not tell me that you had found it? What are you doing with my papers, sir?'

The mischief was out! We tried to explain that the world ought not to be deprived of that which would benefit mankind; that the peace and prosperity of the country might be sacrificed if it were deprived of these discoveries of science, which were calculated to secure such beneficial results.

At length we gained our point, and obtained the full sanction of the late Lady Professor of Girtham College to publish her papers. Thus her obedient pupil is enabled to repay his late instructress for all her kindness to him, and in some measure to compensate the scientific and political world for the loss of one of its most original investigators in the regions of polemical studies, which, not without a struggle, she resigned when she deigned to become his wife.

THE END.
Elliot Stock, Paternoster Row, London.
1886

www.ingramcontent.com/pod-product-compliance
Lightning Source LLC
Chambersburg PA
CBHW070449170726
48291CB00005B/1668